VII and VIII International Conferences on Physics and Technology of Nanoheterostructure Microwave Electronics: Mokerov Readings 2016 & Mokerov Readings 2017

IOP Conference Series: Materials Science and Engineering
Volumle 475

Moscow, Russia
25 May 2016 and 24 May 2017

Editors:

Nikolay Kargin
Vladimir Kulbachinskii
Ivan Vasil'evskii
Alexandr Vinichenko
Dmitry Ponomarev
Rustam Khabibullin
Galina Martirosova

ISBN: 978-1-5108-8218-8
ISSN: 1757-8981

Printed from e-media with permission by:

Curran Associates, Inc.
57 Morehouse Lane
Red Hook, NY 12571

Some format issues inherent in the e-media version may also appear in this print version.

Printed by Curran Associates, Inc. (2019)

For permission requests, please contact the Institute of Physics at the address below.

Institute of Physics
Dirac House, Temple Back
Bristol BS1 6BE UK

Phone: 44 1 17 929 7481
Fax: 44 1 17 920 0979

techtracking@iop.org

Additional copies of this publication are available from:

Curran Associates, Inc.
57 Morehouse Lane
Red Hook, NY 12571 USA
Phone: 845-758-0400
Fax: 845-758-2633
Email: curran@proceedings.com
Web: www.proceedings.com

Table of contents

Volume 475

VII and VIII International Conferences on Physics and Technology of Nanoheterostructure Microwave Electronics: Mokerov Readings 2016 & Mokerov Readings 2017

25 May 2016, Moscow, Russian Federation

Accepted papers received: 21 December 2018
Published online: 18 February 2019

Preface

Papers

IOP Conf. Series: Materials Science and Engineering **475** (2019) 011001 doi:10.1088/1757-899X/475/1/011001

Mokerov Readings 2016 & Mokerov Readings 2017

25 May 2016 & 24 May 2017 Moscow, Russian Federation

The International Conference on Physics and Technology of Nanoheterostructure Microwave Electronics (Open Readings named after Russian Academy of Science Corresponding Member, Professor V.G. Mokerov) is regular annual forum devoted to wide range of modern problems in the field of heterostructure microwave electronics, functional electronics and related topics. Traditionaly the Conference is held at National Research Nuclear University (Moscow Engineering Physics University MEPhI), Moscow, Russia.

The Conference speakers represent various organizations: the Institutes of the Russian Academy of Sciences, the Industrial institutions, the Universities. The presentations reflect the modern state and development of microwave electronics based on AIIIBV nanoheterostructures, SiC and Si nanostructures and other advanced functional materials. Abstracts are combined into thematic sections corresponding to the conference program. The sections are: Fundamentals of nanoheterostructure microwave electronics; Semiconductor microwave devices: fabrication, technology and performance Advanced materials, Heterostructures and superlattices, two-dimensional, one-dimensional and zero-dimensional structures; Functional electronics and Optoelectronics, Terahertz photonics.
The Proceedings contribution papers reflects important novel results concerning fundamental physics of low dimensional structures, functional clusters, physics and technology of heterostructure formation, device application.

Conference Proceedings combined the most significant contributions from the two years of the Conference (2016 and 2017) that have been selected by Conference Program Committee and peer-reviewed.

Guest Editors,

Nikolay Kargin,
Ph.D., vice-Rector, Professor,
National Research Nuclear University MEPhI (Moscow Engineering Physics Institute),
Kashirskoe ave. 31, Moscow, 115409, Russian Federation
NIKargin@mephi.ru

Vladimir Kulbachinskii
Ph.D, Professor,
Lomonosov Moscow State University, GSP-1, Leninskie gory, Moscow, 119991, Russian Federation
kulb@mig.phys.msu.ru

Ivan Vasil’evskii
Ph.D, Assistant Professor,
National Research Nuclear University MEPhI (Moscow Engineering Physics Institute),
Kashirskoe ave. 31, Moscow, 115409, Russian Federation
ivasilevskii@mail.ru

Alexandr Vinichenko
Ph.D, Senior-Researcher,
National Research Nuclear University MEPhI (Moscow Engineering Physics Institute),
Kashirskoe ave. 31, Moscow, 115409, Russian Federation

vanaxel@gmail.com
ANVinichenko@mephi.ru

Dmitry Ponomarev
Ph.D, Deputy Director for Science,
Mokerov's Institute of Ultra High Frequency Semiconductor Electronics of Russian Academy of Sciences (IUUFSE RAS), Nagornyi proezd, 7, Moscow, 117105, Russian Federation
ponomarev@isvch.ru

Rustam Khabibullin,
Ph.D, Scientific Secretary, Leading Researcher,
Mokerov's Institute of Ultra High Frequency Semiconductor Electronics of Russian Academy of Sciences (IUUFSE RAS), Nagornyi proezd, 7, Moscow, 117105, Russian Federation
khabibullin@isvch.ru

Galina Martirosova
Assistant vice-Rector,
National Research Nuclear University MEPhI (Moscow Engineering Physics Institute), Kashirskoe ave. 31, Moscow, 115409, Russian Federation
mokerov@yandex.ru
GRMartirisova@mephi.ru

WEB-sites: http://mokerov.mephi.ru/2018/index_en.html
http://www.mokerov.ru/

Peer review statement

All papers published in this volume of *IOP Conference Series: Materials Science and Engineering* have been peer reviewed through processes administered by the proceedings Editors. Reviews were conducted by expert referees to the professional and scientific standards expected of a proceedings journal published by IOP Publishing.

IOP Conf. Series: Materials Science and Engineering **475** (2019) 012001 doi:10.1088/1757-899X/475/1/012001

Low temperature radiation response of SiGe HBTs

A S Bakerenkov[1], V A Felitsyn[1], A S Rodin[1], Yu D Bursian[1] and V S Pershenkov[1]

[1]National Research Nuclear University MEPhI (Moscow Engineering Physics Institute), Kashirskoe shosse 31, 115409 Moscow, Russian Federation

Corresponding author's e-mail address: ASBakerenkov@mephi.ru

Abstract. Radiation degradation rate of base current in SiGe HBTs was experimentally investigated using X-ray irradiation source with Cu anode at room and low temperatures. The dependences of base and collector current on the emitter-base voltage of the transistors were measured during radiation impact and presented for different total dose levels and irradiation conditions.

1. Introduction

SiGe HBTs are widely used in modern high-frequency electronic devices due to low cost and compatibility with convention silicon technology. There a lot of different applications of high-frequency devices based on SiGe transistors. Excellent balance of electrical performance and compatibility properties of SiGe technology provides a wide range of possibilities for applications of SiGe devices in special operation conditions such as front-end-electronics of particle accelerators or data communication equipment of space-crafts. The common property of both cases is radiation impact during operation. For accelerator applications, it is high energy slowing-down radiation [1]. In the space environment, the main reason of radiation damage of electronics is the impact of protons and electrons from radiation belts of the Earth [2], which can be simulated in laboratory conditions by Co-60 or X-ray sources [3].

As in conventional silicon devices, the radiation impact leads to accumulation of interface traps at the interface of chip substrate and a passivation layer in SiGe transistors. The interface traps act as recombination centers and increases surface component of base current [4]. The increase of base current reduces current gain of the transistors and leads to parametric and functional failures of semiconductor microelectronic devices.

Unlike accelerator applications, the operation temperature of electronic devices during space mission varies in a wide range. The most typical for temperature values for space applications are near the low limit of the operation temperature range. It was experimentally obtained, that low-temperature irradiation reduces the degradation rate of conventional silicon devices [5]. Since it increases failure total dose level, we can consider this feature as very useful. The main purpose of this work is to investigate the degradation rate in SiGe transistors for room and low temperature irradiation.

2. Experimental equipment

As in conventional silicone devices, main electrical performance characteristics of SiGe transistors are the dependences of base and collector current on emitter-base voltage at a fixed bias of the collector junction. Using this data, we can determine the current gain for any emitter junction bias. To perform set different values of emitter junction bias and to measure base and collector currents during irradiation an experimental installation was used. The structure diagram of the installation is presented in Fig.1.

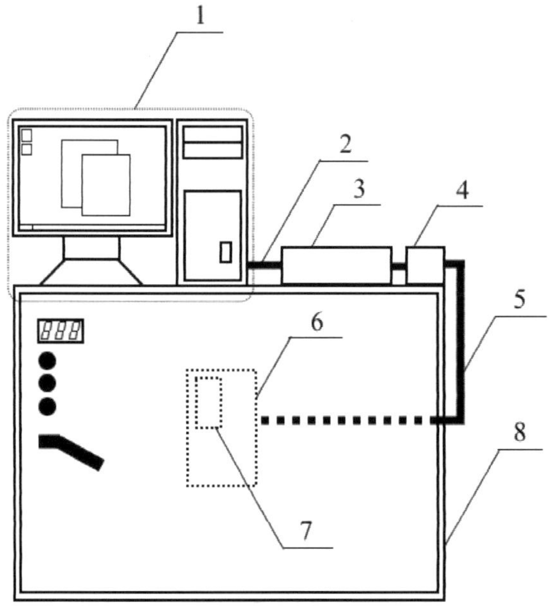

Figure 1. Structure diagram of the experimental installation.

To perform measurements during irradiation, a SiGe-transistor is soldered into printed circuit board (7) with a resistive temperature sensor. The board (7) is in good thermal contact with the active surface of temperature control device (6), the base of thermoelectric modules (Peltier elements). The temperature control device and the board with SiGe-transistor are located in X-ray chamber (8), which involves X-ray source with Cu-anode. The energy of the X-ray beam is 8 keV. The temperature control device and the board with SiGe-transistor are connected with measure (3) and control (4) devices respectively by cable (5). General control and the user interface are provided by the computer (1) through cable (2).

The experimental installation has followed performance parameters:

- available voltage range: -10.0 V ... 10.0 V
- absolute voltage accuracy: ±1.0 mV
- available current range: ±0.1 nA ... ±100 mA
- relative current accuracy: ±1%
- available temperature control and stabilization range: -40^{0}C ... +125^{0}C
- absolute temperature stabilization accuracy: ± 0.1^{0}C
- available range of irradiation dose rates: $1.0 \cdot 10^{-3}$ rad(Si)/s ... 50 rad(Si)/s

3. Experimental results and discussion

For experimental investigations, a sample of commercial available BFU768F SiGe transistor was selected. The BFU768F are NPN wideband silicon germanium radiofrequency bipolar transistors manufactured by NXP Semiconductor in the plastic SOT343F package.

The dependences of base and collector currents of the transistor were measured at various emitter-base voltage values and zero collector junction bias. The dependence of current gain on emitter junction bias was calculated using measured data. The measurements were performed at (25.0±0.1)^{0}C. The irradiation of the sample was performed at 10 rad(Si)/s by X-ray source with Cu-anode in three steps. During the first step, the temperature of the sample was 25^{0}C. The second and the third irradiation steps were performed at -40^{0}C. After each step, the measurements were performed at

(25.0±0.1)^{0}C. The dependences of base current and collector current on total dose are presented in Fig.2 during the first step, after the second step and after the third step of the irradiation. The dependence of base and collector current measured at (25.0±0.1)^{0}C on emitter-base voltage before irradiation and after each irradiation step and corresponding current gain data are presented in Fig.3.

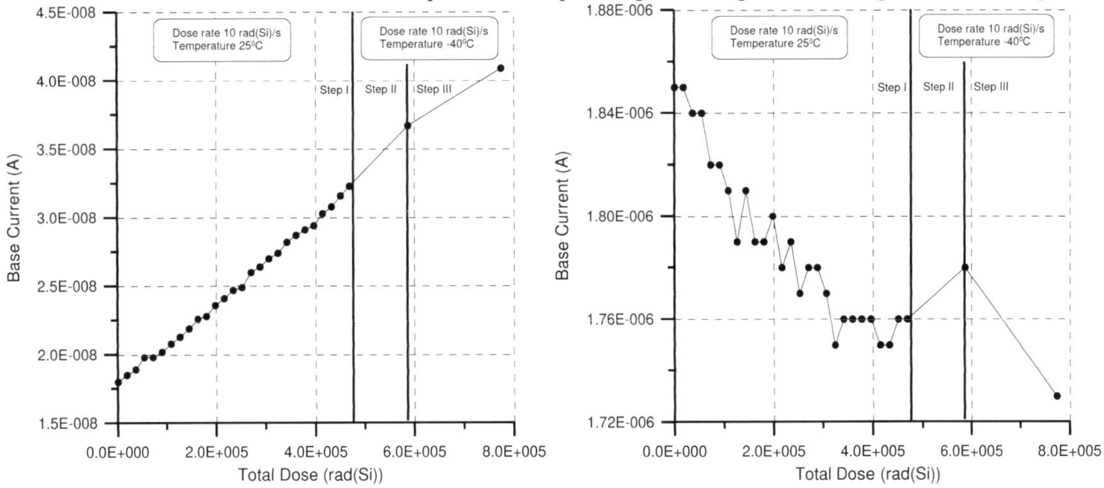

a) b)

Figure 2. Dependence of base current on total dose during irradiation steps (a); Dependence of collector current on total dose during irradiation steps for BFU768F SiGe-transistor.

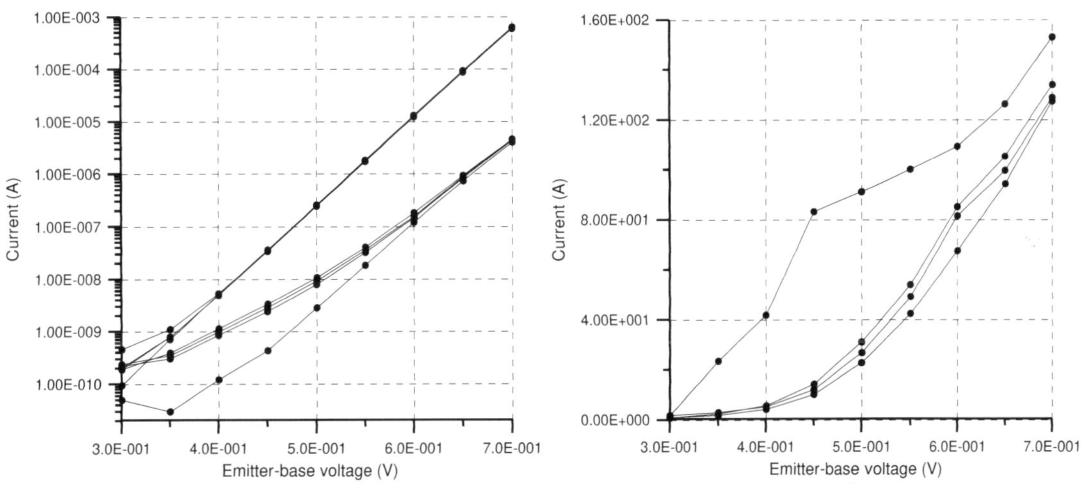

a) b)

Figure 3. Dependence of base and collector current measured at (25.0±0.1)^{0}C on emitter-base voltage before irradiation and after each irradiation step (a) and corresponding current gain data (b) for BFU768F SiGe-transistor.

From Fig.2 and Fig.3 it is clear, that, as in conventional silicon transistors, current gain radiation degradation is connected with radiation-induced an increase of base current because collector current doesn't depend on the total dose significantly for all the emitter-base voltage range. From Fig.4 we can see, that, unlike conventional silicone transistors, the decreasing of the irradiation temperature down to -40^{0}C doesn't lead to decreasing of the radiation degradation rate. It can be explained by decreased oxide thickness over emitter-base junction in comparison with conventional silicon structure. Typical cross section of the SiGe bipolar structure is presented in Fig.4.

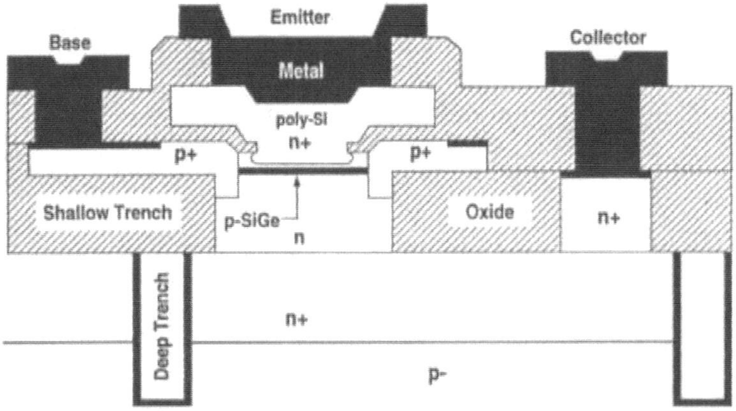

Figure 4. Typical cross section of SiGe bipolar structure [1].

4. Conclusion

As in conventional silicone transistors, the main reason of current gain degradation during radiation impact is increasing of surface component of the base current. Unlike silicon devices, the decreasing of irradiation temperature doesn't decrease the rate of the radiation degradation of SiGe bipolar transistors. It can be explained by decreased oxide thickness over emitter-base junction in comparison with conventional silicon structure. To affect the radiation degradation rate of SiGe device, it is necessary to decrease irradiation temperature down to the low limit of operation temperature range.

5. References

[1] Cressler J D 2013 *IEEE Trans. Nucl. Sci.* **60** 3
[2] Pease R L, Schrimpf R D, Fleetwood D M 2013 *IEEE Trans. Nucl. Sci.* **56** 4
[3] Kessarinskiy L N, Boychenko D V, Petrov A G et al 2014 *IEEE Trans. Nucl. Sci.* **52** 2
[4] Pershenkov V S, Savchenkov D V et al 2009 *Russian Microelectronics* **38** 1
[5] Adell P C, Esqueda I S, Barnaby H J, Rax B, Johnston A H 2012 *IEEE Trans. Nucl. Sci.* **59** 6

Modeling of silicon-germanium heterojunction bipolar transistors

A S Bakerenkov[1], V A Felitsyn[1], A S Rodin[1] and Yu D Bursian[1]
[1]National Research Nuclear University MEPhI (MEPhI), Kashirskoe shosse 31, Moscow, 115409, Russian Federation

Corresponding author's e-mail address: ASBakerenkov@mephi.ru

Abstract. The results of measurements of current-voltage characteristics of SiGe transistors for different temperatures are presented. The extraction results of parameters of test structures of the silicon-germanium SiGe bipolar transistors are presented.

1. Introduction

Nowadays, silicon-germanium heterojunction bipolar transistors have a wide range of applications: microwave devices (low-noise amplifiers, voltage controlled generators, and receiver transmitter devices), ADC, DAC, frequency synthesizers, filters. The SiGe-technology can provide a significant contribution to the electronic component base, which operation does not depend on the environmental conditions [1-3].

The silicon-germanium heterojunction bipolar transistors have several advantages over conventional silicon transistors:

- low temperature dependence of the electrical parameters;
- high speed (cut-off frequency of 210 GHz, IBM, 2005. [4]);
- low noise level (0.6 dB at a gain 20.5 dB and a frequency of 3 GHz);
- increased current gain in the same design size.

These advantages do not lead to large changes in the production technological process of the transistors at the transition from the production of conventional silicon transistor to the silicon-germanium heterojunction bipolar transistors. Therefore production cost of silicon-germanium heterojunction bipolar transistors is almost the same as the conventional silicon transistors. In the process of designing integrated circuits (ICs) on the basis of the silicon-germanium heterojunction bipolar transistors, it is possible to use standard libraries and design methods.

For the development of microwave electronic devices based on silicon-germanium heterojunction technology, it is necessary to simulate the operation of this device at the design stage. Spice models of the electronic components which are part of the microwave devices are used in the development of complex models of microwave devices. One of these models is the model of SiGe heterojunction bipolar transistor, which is based on the model of Hummel-Pune. The purpose of this article is to extract the parameters of the model for different transistor operation temperatures. For the extraction of the parameters, test structures of the silicon-germanium heterojunction bipolar transistors are used.

2. Description of samples SiGe bipolar transistors and testing

Research object is the high-frequency bipolar npn-transistors, which is produced using the BiCMOS technological process with design rules of 0.25 microns.

Transistor chip consists of two (SGB25V_016P) and four parallel-connected transistors npnVh (SGB25V_019P). The transistors with the size of the emitter 3.36 x 0.42 mm^2 are used. One transistor includes 16 emitters (the grid of emitters 8x2). Transistor chip includes the polysilicon resistors 3 kOm.

The structure which consists of two the SiGe heterojunction bipolar transistors is shown in Fig.1. Base, emitter, and collector regions are marked correspondingly «b», «e» and «c». The connecting circuit of the SiGe transistor chip with a pin of the microchip package is shown in Fig.2a. Photo of the finished SiGe transistor sample is shown in Fig.2b.

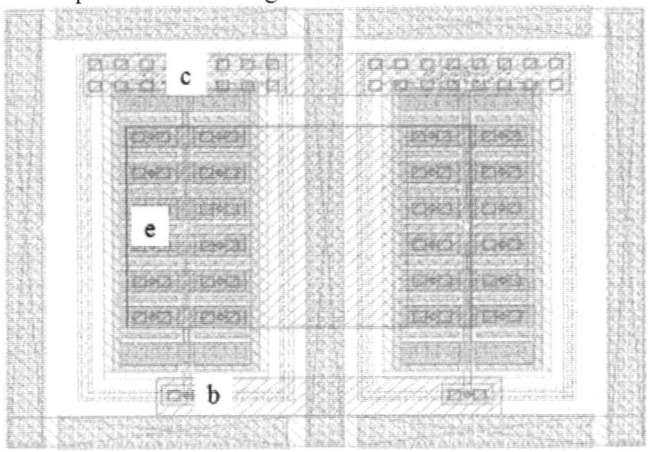

Figure 1. Topology of SiGe heterojunction bipolar transistor.

Fig. 2a shows the chip topology: large octagons – contact pads which are welded to the metal pins of the chip to the microchip package. Metallization is marked by dashed lines of orange and blue colors, polysilicon resistors – purple stripes with dots. Transistor blocks are located in the center. To increase the reliability and percentage production quality sample, base, emitter and collector pins is duplicated.

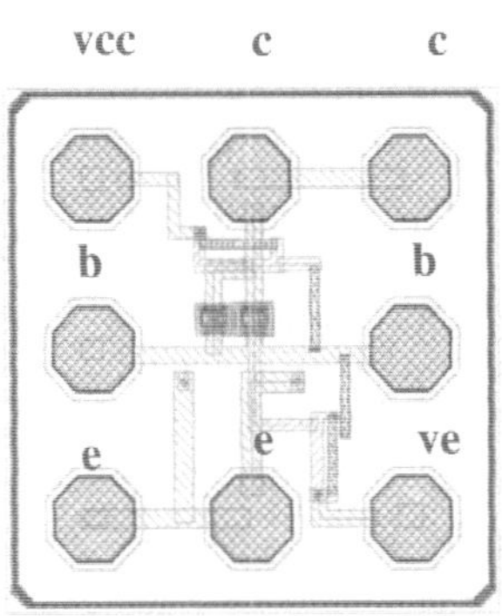

(a)

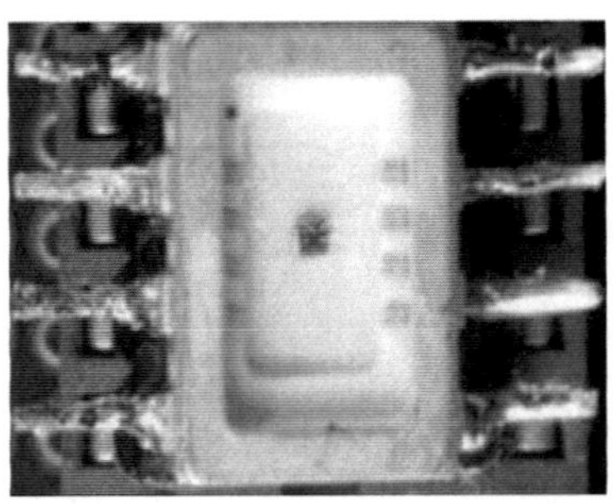

(b)

Figure 2. The topology block of two transistors (a) and photo IC chip (b).

For the testing, these samples were used the measuring system and heat chamber. The measuring system consists of PC with the connected device in which integrated voltage sources and ammeters. The measuring system enables to measure a current-voltage characteristics (CVC) of the samples. The heat chamber enables to set the temperature of the sample at which the measured current-voltage characteristics.

3. Model of silicon-germanium heterojunction bipolar transistors

The Gummel–Poon bipolar transistor model for circuit simulation in LTSpice VI was used. This model has many parameters, most of which is used to set the frequency characteristics of the transistor. Because of in this paper static characteristics of bipolar devices are used, the circuit simulation is not required to set frequency parameter values. Part of the model parameters used to describe the transistor in reverse mode. Since, in operational amplifiers, transistors are used in the forward mode, the values of these parameters do not affect the simulation results.

Using this assumption, for base and collector current-voltage characteristics of bipolar transistor it's possible to write relationships:

$$I_c = I_s \exp(U_{eb}/\varphi_T), \tag{1}$$

$$I_b = (I_s/B_f) \exp(U_{eb}/\varphi_T) + I_{se} \exp(U_{eb} \tag{2}$$

in which I_c – collector current, I_b – base current, U_{eb} – emitter-base voltage, I_s, B_f, I_{se} и n_e – Gummel–Poon model parameters. Since the radiation degradation of the transistor is determined by the increase in the base current at constant collector current, it possible to assume that the parameter I_s does not change at radiation impact. The value of B_f defines pre-exponential multiplier of ideal base current component (proportional $\exp(U_{eb}/\varphi_T)$). This component of the base current is the sum of the currents in the emitter injection and recombination in the active base region. Since in modern transistor size of the active base region is small enough, recombination current in the active base region is much smaller than the current injected into the emitter.

4. Extraction parameters

As a result, CVC (Fig. 3) and the dependence of the static gain on the base-collector voltage for different temperatures for the case of the active and inverse mode are measured.

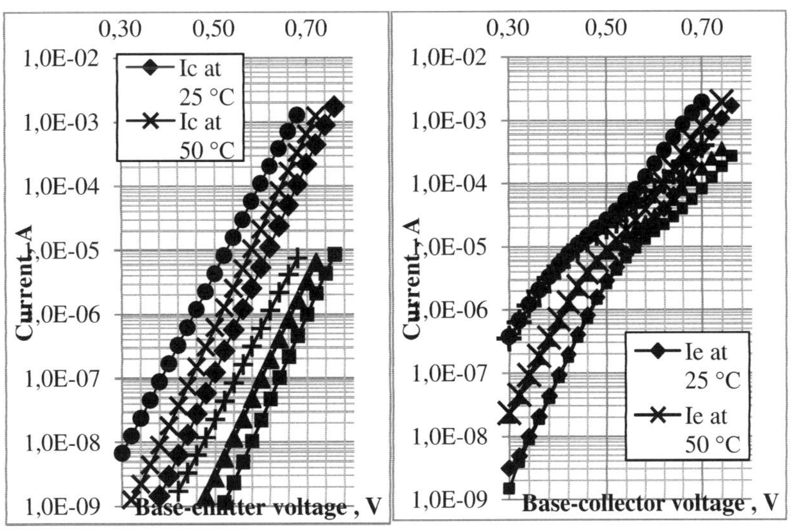

(a) (b)

Figure 3. Hummel plots for the different temperatures (a - normal active mode, b - inverse mode).

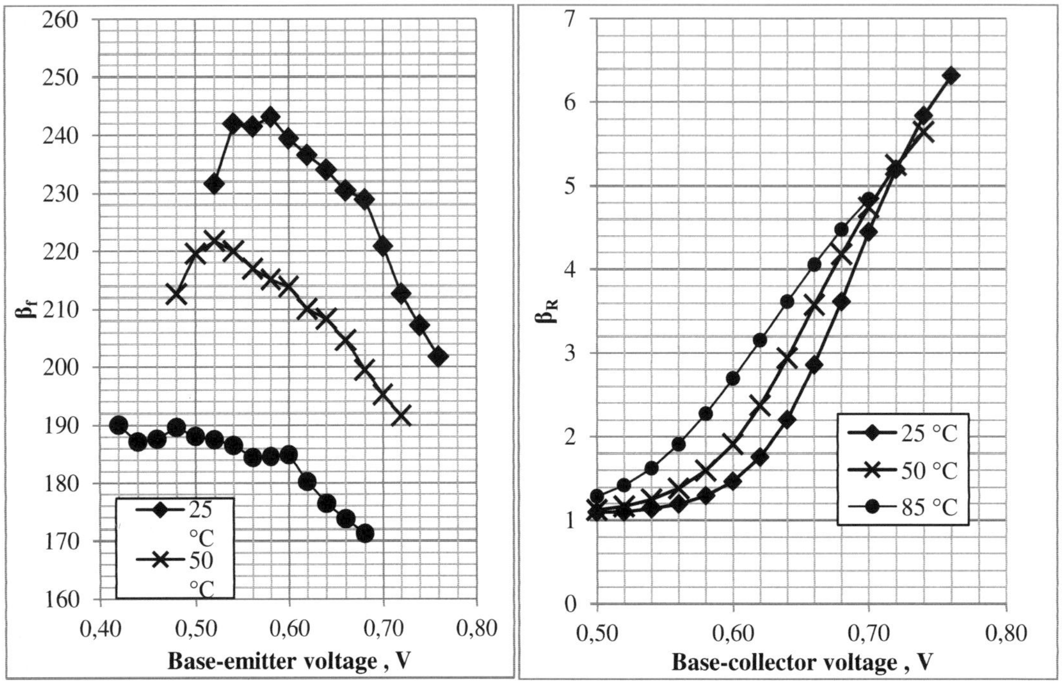

(a) (b)

Figure 4. The dependence of the static gain on the base-emitter voltage (normal mode) (a) and the dependence of the static gain on the base-collector voltage (inverse mode) (b).

Extraction of the model parameters was performed by using the curves presented in Fig.4. Parameters obtained by extraction procedure are shown in Table 1.

Table 1. The parameters of the model SiGe heterojunction bipolar transistor extracted from experimental data.

T, °C	I_s	B_f	I_{se}	n_e
25	$1.0 \cdot 10^{-15}$	243	0	1
50	$2.0 \cdot 10^{-14}$	221	0	1
85	$5.0 \cdot 10^{-13}$	190	0	1

5. Conclusion

This article presents the results of measurements of the current-voltage characteristics of test structures of the SiGe bipolar transistor for different temperatures. The result of parameters extraction of test structures of the SiGe bipolar transistor, which can be used for schematics simulation during high frequency developing process is presented.

6. References

[1] Cressler J D 2005 *IEEE Trans. Nucl. Sci.* **93** 1559
[2] Cressler J D 2005 *Proc. IEEE Bipolar/BiCMOS Circuits and Technol. Meeting.* 248
[3] Cressler J D 2006 *Proc. IEEE Silicon Monolithic Integrated Circuits in RF Systems.* 3

Formation of synergetic effects in integrated circuits operational amplifiers based on the current dynamic cascades

A A Lebedev[1], I I Liakhchilina[1] and N M Klokov[1]
[1]National Research Nuclear University MEPhI (Moscow Engineering Physics Institute), Kashirskoe shosse 31, 115409 Moscow, Russian Federation

Corresponding author e-mail address: Lebedev741@mail.ru

Abstract. In this paper, the existing IC developments (Bazes operational amplifier with an automatic set of modal current, current feedback operational amplifier (CFA), a two-channel operational amplifier with an accurate combination of AFC areas of low and high-frequency channels) from the standpoint of synergistic effect is analyzed. New architectural, structural and circuit solutions to improve the accuracy, speed and radiation characteristics with the help of directional formation of synergistic effect (SE) proposed by the authors are also considered. A number of criteria to ensure future product improvement compared to commercially available are presented.

1. Introduction

Currently, there are several major developments in the creation (design) of industrial ICs wide application [1-4]. These include the following developments: a high-speed current feedback operational amplifier (op amp) (CFA), which successfully solved the problem of the relative independence of the gain and bandwidth of negative feedback amplifier (NFA) [5]; wideband amplifier [1] with automatic modal current in CMOS technology, as well as wideband op amp with accurate combination Bode diagram plots of low-frequency and high-frequency amplifier stages.

The new approach connected to use of controlled macrocells of current cascades [6] in order to expand their functionality and to use of local negative feedback, proposed in [5].

2. The presence of synergy effects, their analysis and use to improve the quality of industrial op amp IC

Improvement of the technical characteristics mentioned in the introduction can be considered (examined) from the perspective of synergistic effect.

In synergy dynamic systems are combined, and the new entity is not equal to the sum of the parts and forms a system of another organization or system level. An indication of this is the amplifier [2] (let's call it Bazes amplifier), in which SEFF is achieved by ring connection of two gain stages. This allows setting a single ring current by automatic tuning the resistances of current sources formed by the diode-connected transistors in the NMOS and PMOS, respectively. Bazes amplifier is a three-stage structure; the first two stages have feedback of -3 dB, for a total of -6 dB. Stability of Bazes amplifier covered by feedback is achieved by inserting local negative feedback. This allows, on the one hand, to expand the bandwidth at a gain of 32 dB and on the other hand, receive the phase margin of 80-85 °.

IOP Conf. Series: Materials Science and Engineering **475** (2019) 012003 doi:10.1088/1757-899X/475/1/012003

3. Methods of improving accuracy and speed parameters of industrial grade LM124 bipolar op amp IC

In this paper, the use of main architectural, structural and circuit solutions, viewed from the perspective of the SE is proposed.

The main difference of proposed solutions is connected to the organization of an improved (compared to conventional) op-amp architecture, in which instead of one differential cascade (DC) a parallel connection of two stages with a common mid-point is used. This midpoint is fed with in-phase component of the output signal, which allows weakening the dependence of the dynamic offset on the impact of the input common-mode signal.

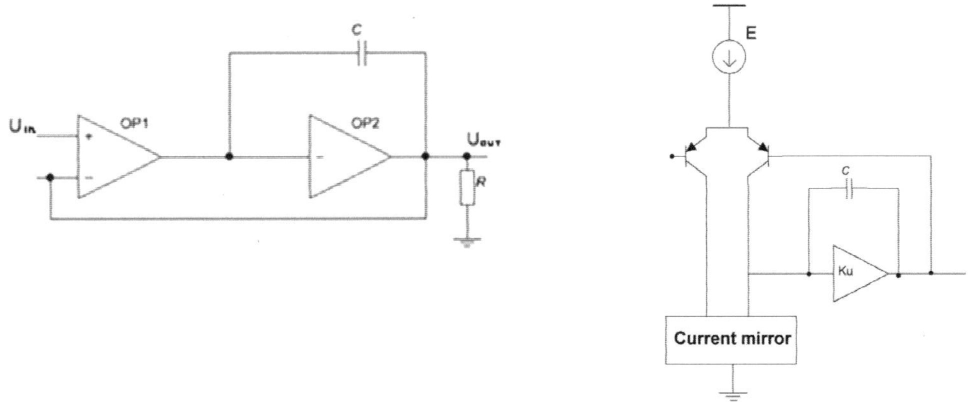

a) b)

Figure1. LM124 op amp: a - a standard LM124 op amp architecture, b - a standard LM124 op amp structural organization.

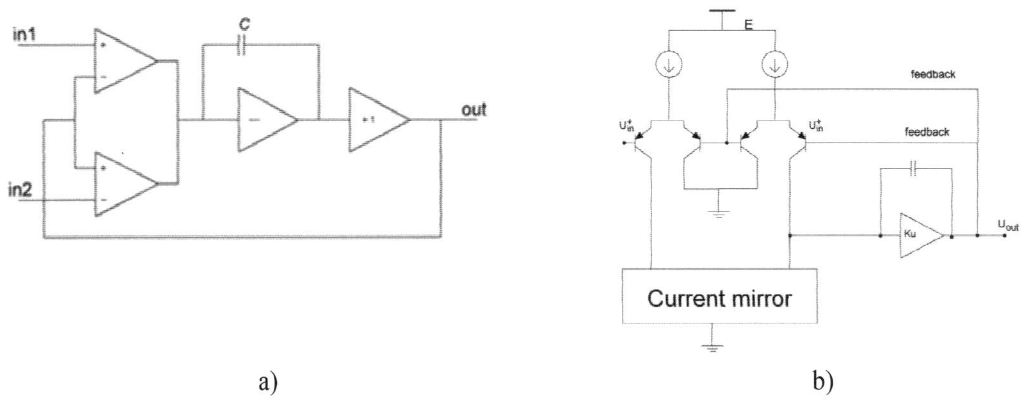

a) b)

Figure2. Proposed op amp with extended bandwidth, increased slew rate and high resistance to HCP and the effect of in-phase signal: a - proposed op-amp architecture, b - proposed op amp structural organization.

Fig. 1 shows a standard LM124 op-amp architecture and structural organization. The disadvantage of this architecture is a hard relationship between accuracy, frequency and speed characteristics. Also, the in-phase component of the signal shift results in an undesirable constant component at the output of op amp. This shift can be described by the following simplified equation:

$$\Delta U_{out} = K_{dd}(\Delta U_{ind} + \frac{\Delta U_{insp}}{K_{ds}}) \quad (1)$$

where ΔU_{ind} - differential component of the input signal ΔU_{insp} - in-phase component of the input signal, K_{dd} - differential signal gain, K_{ds} - input to differential transmission coefficient. It can be shown [3] that the inverse of common mode rejection ratio (CMRR) for the typical architecture of the op amp is evaluated using the following expression:

$$\frac{1}{CMRR} = \Delta\mu = \mu_m \frac{\Delta\mu}{\mu_m} \tag{2}$$

Where
$$\mu_m = \frac{1}{W}\left(\frac{dW}{dU_{dd}}\right) = 5 \cdot 10^{-4} \tag{3}$$

For the proposed architecture in Fig. 2 for AC common-mode the following relation [3].

$$\frac{\Delta\mu}{\mu_m} = \frac{\Delta U_{dd}}{U_{dd}} \tag{4}$$

Then CMRR = 5×10^5. Thus, the proposed architecture of the input DC allows to increase the CMRR on order and weakens the rigid relationship between static and dynamic zero offset voltage, and also to reduce the response to the impact of HCP by doubling the current signal supplied to the input of the intermediate stage.

4. Dynamic current stages and the formation of synergistic effect based on them

In this section the dynamic current cascades are discussed, such as current mirror (CM) with automatic magnification (threw) the modal current while the degradation of the gain current β (further - Radiation-resistant reflector current Radiation Hardness Current Mirror (RHCM)), as well as a two-tier current reflector (current mirror) with a current collector to increase the dynamic output resistance and hence voltage gain.

Radiation-resistant current reflector shown in Fig. 6. Formed by including the amplifier stage arranged in a common base between points 1 and 2 of. In [5] the principles of analog circuits by replacing the one-loop current reflector structure by a two-tier structure Fig. 3 are discussed, which allows implementing the function:

$$I_w = \sqrt{\left(I_x^2 - I_y^2\right)} \tag{5}$$

An indispensable condition for the proper functioning is supply of variable sources of current I_x, I_y to a low-impedance inputs. This function was derived using translinear principle to the existing CM applications.

Returning to Widlar CM automatic increase of modal current at a current gain degradation β was tasked. Such formulation of the problem requires the inclusion of an amplifier stage with a common base and installation of the AC voltage on the basis of zero in the gap between points 1 and 2. The complete circuit is shown in Fig. 6, and its characteristic output current dependence on β in Fig. 5 In the resulting circuit, transmission factor is K_u:

$$K_u = 1 + \frac{\beta - 2}{\beta^2 + 2} \qquad 6)$$

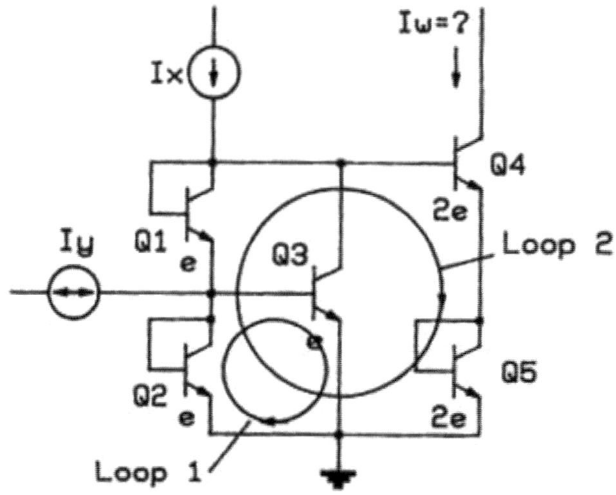

Figure 3. Two-loop structure CM shown by Toumazou.

From the diagram, it follows that the inclusion of a common base stage between points 1 and 2 produces a local positive feedback wherein unity loop gain is less than 1 which provides circuit stability when used as an input stage, either as an active collector load. Thus, the structure organization of the op amp on the basis of the proposed radiation-resistant CM reduces shift of the operating point at the output and thus extend the life cycle of a spacecraft in Earth orbit for 1 - 2 years.

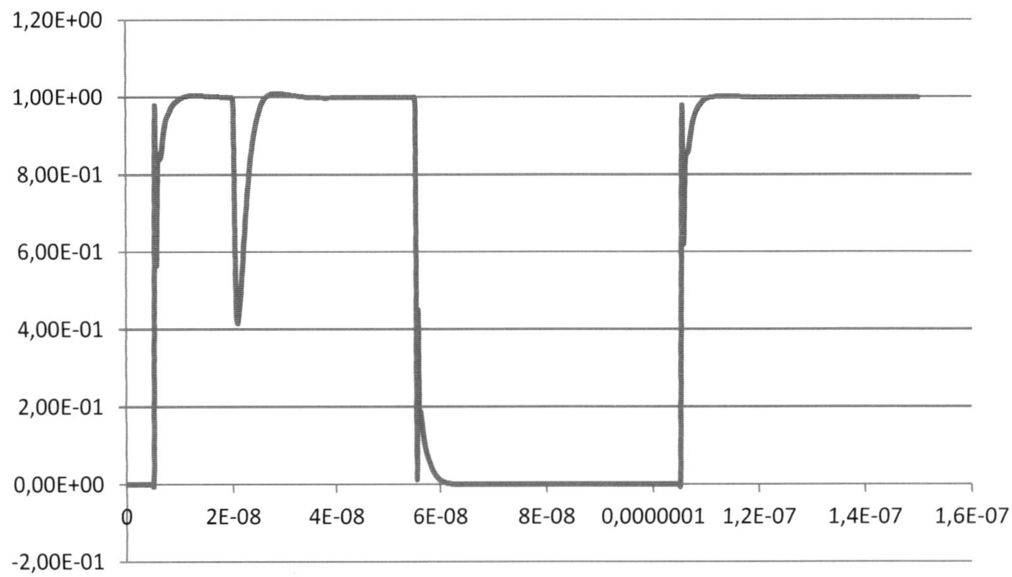

Figure 4. Transient Response to HCP.

As a result, the obtained dependence reflects the improvement of scheme's characteristics: no emissions on the transient response, reducing SET to 3 ns while reducing the amplitude of the HCP reaction.

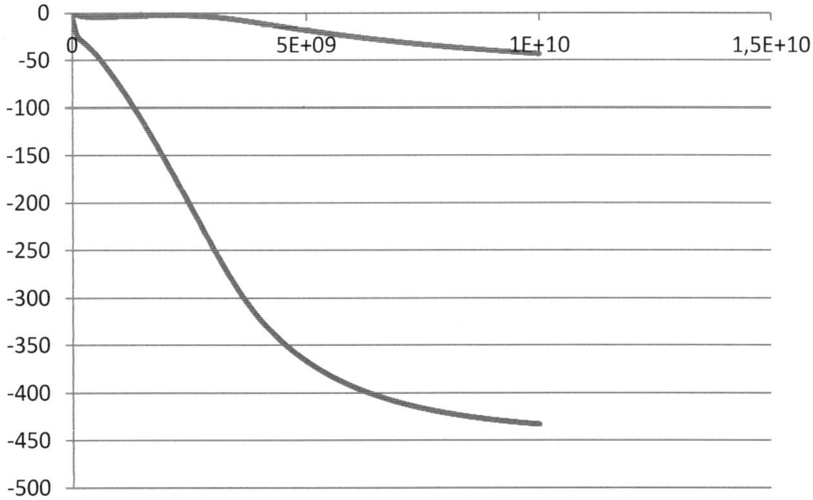

Figure 5. Frequency response (phase response and frequency response).

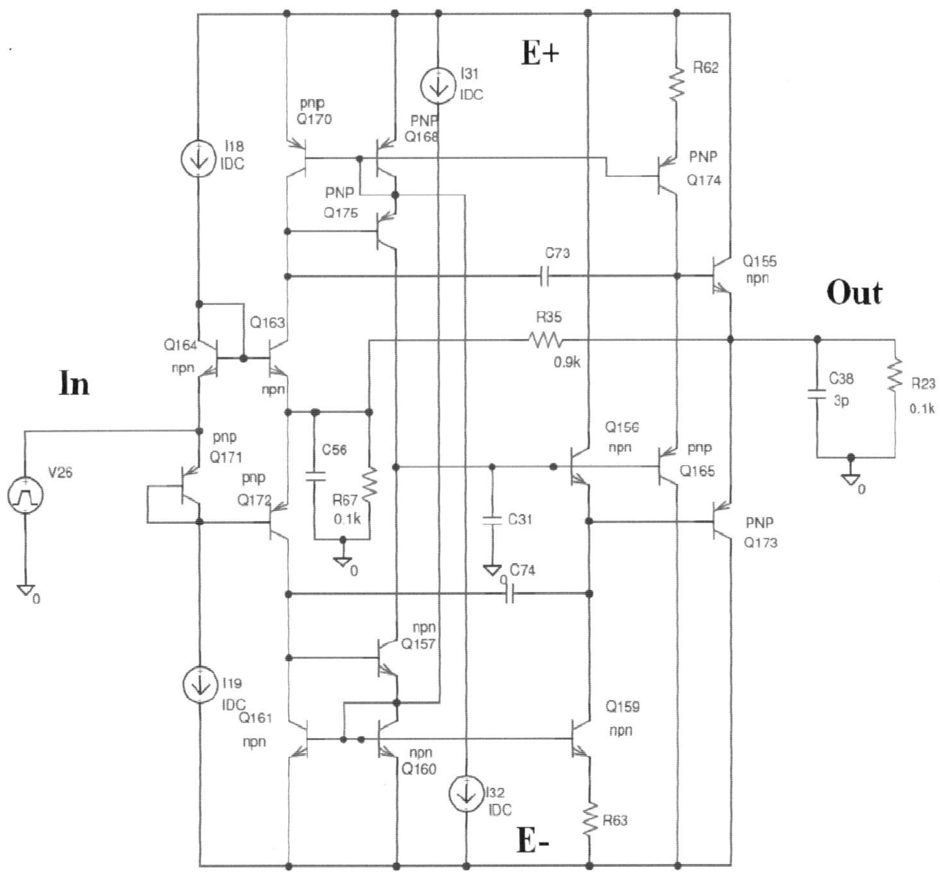

Figure 6. A simplified scheme of developed op amp with current feedback.

5. Discussion

It can be affirmed that improving accuracy, frequency and radiation characteristics is directly related to the formation of synergistic effect (SE), which self-formation can be described by the following criteria:

1. The presence of a local positive feedback, providing structure self-organizing in the predetermined direction to forming of SE.
2. Insertion of the dynamic current element (cascade) controlled by feedback.
3. The ability to control the system as a whole using negative feedback.
4. Maintaining stability of an op amp as long as the transfer constant loop gain less than one.
5. The openness of the system and communication with the external environment to maintain the energy of self-formation.

Thus, it is possible to predict the characteristics and parameters of the designed product.

References

[1] Roche N J-H, Perez S, Mekki J et al 2011 *IEEE Trans. Nucl. Sci.* **58** 2980
[2] Bazes M 1991 *IEEE JSSC* **26** 165
[3] Soclof S 1985 *Analog Integrated Circuits (Los Angeles: Prentice-Hall)*
[4] Lebedev A A, Nazarenko A E., Vakhnenko M S et al *2016 IOP Conf. Series: Mat. Sci. Eng.* **151** 012038
[5] Toumazou C, Lidgey F J, Haigh D G 1990 *Analogue IC design: the current-mode approach*, (London: Six Hills Way) 23
[6] Lebedev A A, Felitsyn V A, Komleva V A and Komlev A A 2016 *Proc. on Int. Federation for Medical and Biological Engineering (Chisinau)*(Springer) **55** 23

Automated unit of the chemical wet etching

Yu V Sukhoroslova[1], D S Veselov[1] and Yu A Voronov[1]

[1]National Research Nuclear University MEPhI (Moscow Engineering Physics Institute), Kashirskoe shosse 31, 115409 Moscow, Russian Federation

Corresponding author's e-mail address: ASBakerenkov@mephi.ru

Abstract. This paper is devoted to the development of wet etching unit and its control system. The experimental model represents the reaction chamber, equipped with the stirrer and steam-gas condenser, located inside the heated chamber. The unit is also equipped with heat-insulated reservoirs for preheating of etchants and water. The control system provides automated supplying of liquids into the reaction chamber. This allows realizing multistage etching in single technological cycle with etchants replacement and substrate washing, without its extraction from the reaction chamber. With using the unit was studied deep anisotropic etching of silicon in various alkaline etchants. Was revealed, that unit provides the uniformity of etched surface on the level of about 1 μm.

1. Introduction

Currently, for the technology of integrated circuits, along with the plasma chemical etching techniques are increasingly applied the liquid silicon etching. It differs with its simplicity, implementation and low cost, also ensures high reproducibility and uniformity of the surface etching. However, it is difficult to stabilize the temperature and concentration of the etchant when the wet etching. This complicates the manufacturing process of sensing elements based on dielectric membrane structures for the gas concentration sensors. A typical construction of such sensing elements is shown in Figure 1.

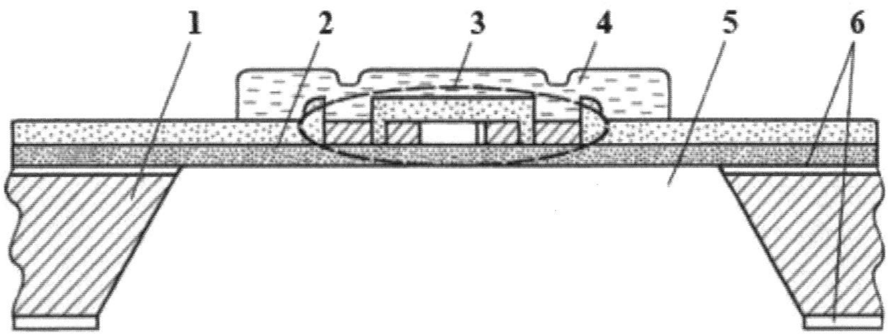

Figure 1. Sensor element based on a dielectric membrane structure: 1 - silicon substrate; 2 - dielectric membrane; 3 - structural elements (resistive heater, a resistive temperature sensor contacts to the sensitive layer); 4 - the sensitive layer; 5 - cavity under the membrane; 6 - a layer of thermal silicon oxide.

In the manufacturing process of such sensing elements cavity is formed under the membrane deep anisotropic etching of silicon in alkaline solutions etchants. Formation of cavities is made in two stages. On the first step is formed the cavity mordants through anisotropic silicon etching to a depth of 80-90% of the substrate thickness. Then carried out the steps of forming the structural elements and the sensitive layer. After that, the final stage is carried through anisotropic etching of silicon substrate with a protection of the front side against contact with the etchant. Depending on the size of the substrate, it is required to provide one-time mordants of cavity under membrane to a depth of 0.3 to 0.8 mm. In the process of the deep etching solution loses its etching properties. This lowers the etching rate and uniformity, which requires periodic replacement of the etchant. Stages of the etchant solution replacement significantly complicate the manufacturing process and reduce the yields of devices, which adversely affects the final cost of the product.

While forming the sensing elements based on dielectric membrane structure is extremely important to obtain simultaneously the entire membrane area of the substrate. This requires providing a high level of silicon etching uniformity. More information about etching process can be found in following article [11]. Otherwise, the etchant can etch one of the membranes and to come under the protective device, resulting in the poisoning of the sensitive layers on a substrate. Furthermore, it is required to provide uniform access of the etchant to the etching surface. Also, in order to avoid destruction of the substrate fixed in the protection device when it is immersed in the etchant is necessary to ensure smooth heating and cooling.

The existing equipment for the wet etching process, as a rule, is designed for the individual process steps for the group of substrates. There is no special technological equipment for the group of technological operations of etching a substrate, which is represented on the market. Therefore, the authors had a task to develop the stable wet etching equipment with an automatic control

2. The construction

To automate the anisotropic etching process the structural design was made in various modifications, and it was empirically determined the optimal device design for the temperature stabilization and the concentration of the etchant shown in Fig. 2 (hereinafter referred to as indicated by Fig. 2). It consists of the following main parts: a heat-resistant heat-insulating casing 6, the outer contour 8 warms up with coolant; inside etching chamber 10 for loading protection device with the one-sided etched 14 (3,4 Picture) with the substrate; condensing the vapor-gas suspension system 25; storage tank 38 with deionized water.

Thermostat performs latitudinal load control with a period of 2.56 seconds and sets the current limit load in a range from 0 to 100% in 1% increments. Empirically were found, that the equipment of heated loop keeps the etchant temperature in the etching chamber with an accuracy of ± 1 ° C, the control of which is held by a thermocouple TCA 13, located in the etch chamber. It is also equipped with an external circuit, which heats the channel 20 for draining the coolant from the built-in water valve 18 and the electric coolant channel 5 for the Gulf.

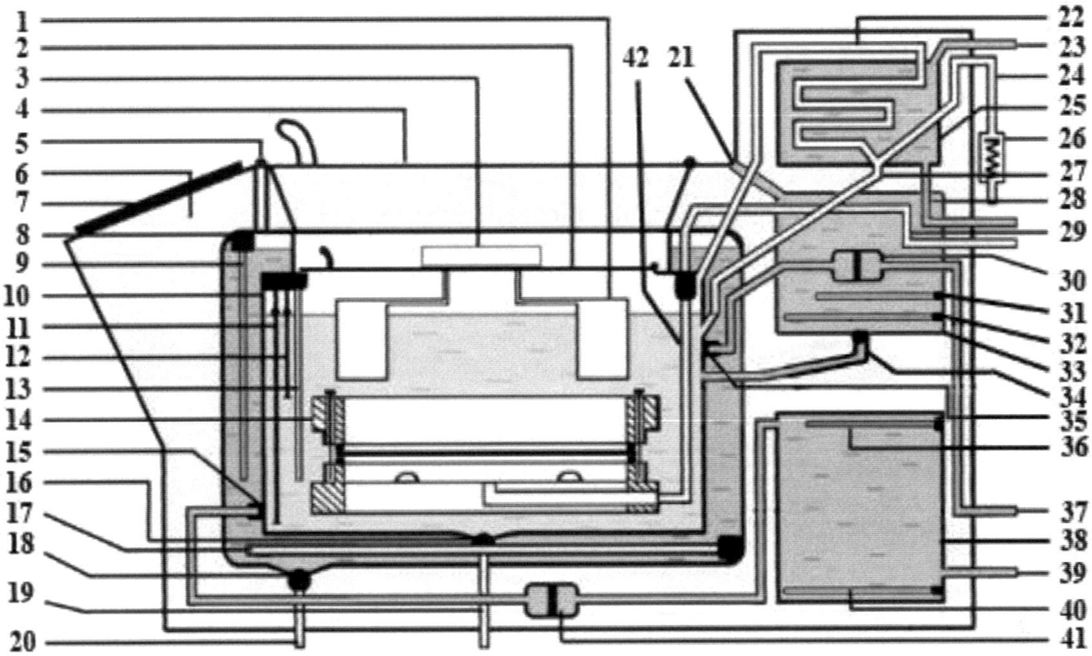

Figure 2. The device construction.

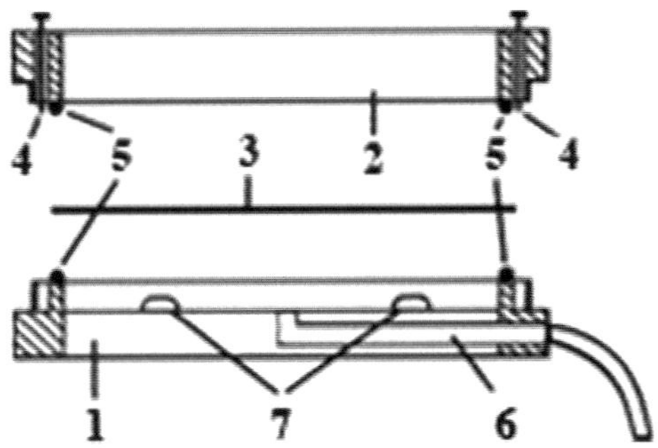

Figure 3. The protection device for the one-sided etching:
1 - the Teflon basis; 2 - Teflon ring; 3 - substrate; 4 – Teflon screw; 5 - chemical and heat-resistant O-ring; 6 - channel for removal of excess pressure; 7 - LED.

Figure 4. A substrate at the security device for the sided etching.

Access to the etching chamber is made through an insulated door 4, and the control of the final stage of the etching process is performed visually through a transparent and sealed door 2, provided with a valve to drain the over-pressure from the etching chamber, at which is the drive of 3 blades for stirring the etchant. Also, the chamber for etching is provided with a channel 19 for the drain of the etchant with built-in water with an electric valve 16 and initially filled with deionized water, which is gradually heated to the temperature of etching. The device is equipped with a capacity of 33 with an opening 21 for filling the etchant, which is heated to a temperature etching using coiled heater 32, the operation of which is controlled by a thermocouple temperature controller TCA 31.

Upon reaching the desired temperature of the etchant solution, the water is drained from the etching chamber and the etchant is then supplied to the chamber through the channel 34, equipped with an electric valve. The deionized water is supplied through the channel 39 to the capacity storage 38. The water temperature is maintained throughout the etching process at a temperature equal to the temperature in the etching chamber with spiral heater 40, the operation of which is controlled by a second thermostat Warta TP410-1 by a thermocouple 36 TCA.

After the etching process, deionized water is supplied through a valve 15 by water pump 41 to the etching chamber. The water supply is controlled by a float liquid level sensor 12. There is a similar water valve 35 on the back side of the camera, through which the etchant is pumped out (washed out) by the pump 30 and the channel 37. The process is controlled by an analog linear float liquid level gauge 11.

The steam-gas suspension is given into the condensation system from the etching chamber through the channel 22. In the condensation system the vapor condenses, and the liquid returns to the etching chamber through the channel 27, and the gas is discharged from the device through the channel 24, equipped with an air check valve 26.

A condensation cooler is fed to the channel 28 (the easiest way is to connect to the cold water) and output through the channel 23. When placing the protective device for one-sided etching in the etching chamber 14, channel 42 is connected to the channel 29 for discharging the excess pressure in the installation. Protection device 14 is equipped with LEDs. On completion of step etching with a residual silicon layer (less than 10 microns) in the cavity for the membrane formation, the membrane LEDs change color, as shown in Figure 5, to stop the etching process. This method of stopping etching is used at the non-applicability of other methods within the framework of the developed technology. The control equipment is placed on a single dashboard 7 for the convenience of work.

Figure 5. A membrane to the light, residual silicon layer: a) less than 10 microns; b) less than 5 microns; c) less than 3 microns.

3. Conclusion

There was developed an automated system for carrying out multi-step process cycles of wet etching with the replacement of solutions and washing the substrate without its removal from the etch chamber. It was revealed that the designed setting provides the etching temperature stability with an accuracy of less than ± 1 ° C in the etching temperature range (50-100) ° C. Also, the installation provides stability of the solution concentration in the etching process with an accuracy of less than ± 1% of the mass fraction of each component of the solution. The unit provides uniform access of etchant to the substrate, as well as a smooth substrate heating and cooling. Experimentally was found that by etching to a depth of several hundred micrometers the unit provides silicon etch uniformity across the substrate area at the level of 1 micron. The installation price is only about 5 thousand dollars. This makes it particularly attractive in the market of technological equipment.

4. References

[1] Nishi Y and Doering R 2007 *Handbook of Semiconductor Manufacturing Technology (New York: Taylor & Francis Group)* 1720
[2] Veselov D S and Voronov Yu A 2016 *J. Phys. Conf.* **747** 012022
[3] Veselov D S, Bakun A D and Voronov Yu A 2016 *J. Phys. Conf.* **748** 012017
[4] Veselov D S, Bakun A D, Voronov Yu A et al 2016 *J. Phys. Conf.* **748** 012015
[5] Rouhi J, Mahmud S, Hutagalung S D et al 2012 *Semicond. Sci. Tech.* **27(6)** 065001
[6] Shikida M, Inagaki N, Sasaki H et al 2010 *J. Micromech. Microeng.* **20(1)** 015038
[7] Veselov D S 2015 *Phys. Procedia* **72** 500
[8] Veselov D S and Voronov Yu A 2015 *Phys. Procedia* **71** 423
[9] Veselov D S, Voronov Yu A and Vanyukhin K D 2015 *Phys. Procedia* **72** 495
[10] Zenkevich A V, Lebedinskii Yu Yu, Matveev Yu A et al 2010 *Russ. Microelectron.* **39(3)** 165
[11] Rakhimov R A, Osipov E V, Dovzhenko D S et al 2016 *J.Phys.: Conf. Series* **737(1)** 012026

IOP Conf. Series: Materials Science and Engineering **475** (2019) 012006 doi:10.1088/1757-899X/475/1/012006

Features of ion mobility spectrometry hardware-software complex

Y R Shaltaeva[1], E A Gromov[1], V V Belyakov[1], A V Golovin[1], V K Vasilyev[1], E K Malkin[1], M A Matusko[1] and I I Ivanov[1]

[1]National Research Nuclear University MEPhI (Moscow Engineering Physics Institute), Kashirskoe shosse 31, 115409 Moscow, Russian Federation

[1]Corresponding author's e-mail address: YRShaltayeva@mephi.ru

Abstract. The ion mobility spectrometer (IMS) is a complex hardware and software device for gaseous analysis. The article describes bipolar IMS control system. The paper lists the main requirements in electronic units, the modes of their operation, the ranges of normal operation and the algorithm for handling emergency situations. The main electronics include: a unit for collecting and processing spectrometric data, an ion source control unit, a gate control unit, high-voltage generators, an ion mobility spectrometer control and synchronization system. The structure of additional electronics includes: control unit of actuating devices, air conditioning system, self-diagnostics and protection system. This article may be useful to designers of complex electronic analytical systems.

1. Introduction

The key task in design of modern inspection equipment is the detection of ultra-low trace amounts of explosive, narcotic, toxic [1-4] and other hazardous substances is. Nowadays there is an active fight against smuggling of drugs and explosives. Moreover, there is a need for simultaneous detection of substances of various classes in a number of tasks [5, 6]. Thus, one of the crucial requirements of modern inspection technology is the simultaneous detection of a wide range of substances.

Currently, following methods are most frequently used for the detection of prohibited substances: chromatography [7, 8], mass spectrometry [9] and ion mobility spectrometry (IMS) and detection by trained dogs [10-15]. The method of ion mobility spectrometry allows you to collect sufficiently complete information about the composition of impurities in a gaseous medium, with relatively small size, equipment weight and energy consumption requirements. On the basis of their characteristics, ion mobility spectrometers are considered the most promising and versatile class of instruments in inspection systems. Therefore, ion mobility spectrometers are widely used in electronic checkpoints in enterprises [16–18].

The principle of IMS is based on the separation of ions by time of flight in a gaseous medium in a constant electric field. The resulting time-of-flight spectrogram, in which the peaks of substances can be distinguished, characterizes the presence of ions with a certain mobility. The magnitudes of the mobility of the ions of the test substance, reduced to standard values of temperature and atmospheric pressure, make it possible to identify the composition from tabular values recorded in the database of substances of the instrument.

The purpose of simultaneous detection of substances of different classes requires joint detection of ions of both positive and negative polarity. Therefore, the task of developing a bipolar high-resolution

ion mobility spectrometer for the express detection of ultra-small quantities of substances and its control system is a topical issue.

2. Experimental equipment

The electrical circuits in the device of the ion mobility spectrometer can be divided into two groups - main and additional electronics. Within the main electronics category the following are included: a collection unit for processing spectrometric data, a unit for the control of the ion source, a gate control unit, high voltage generators, a control and synchronization system of the ion mobility spectrometer. The category of additional electronics includes the following: a control unit for executive devices, an air conditioning system, as well as a self-diagnosis and protection system.

The synchronization system provides the generation of pulses with a given time structure within the scope of the spectrometric cycle. The sequence of events is shown in Figure 1.

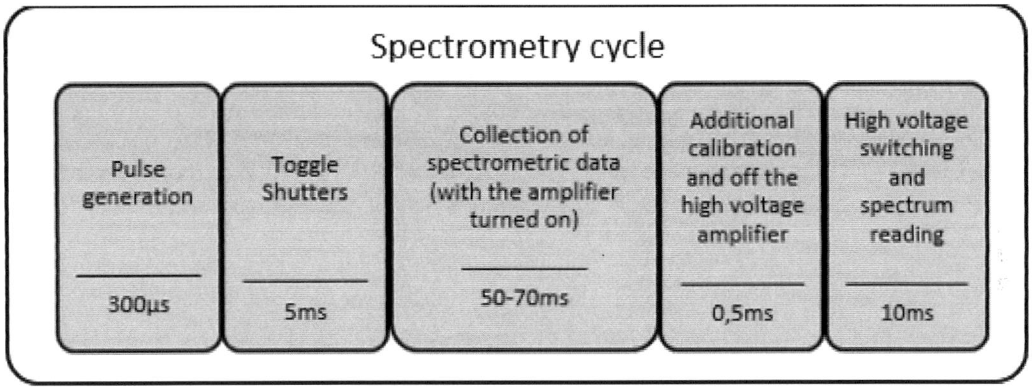

Figure 1. Structure of the spectrometric cycle.

The spectrometric cycle begins with the process of generating corona pulses, which continue for 300 μs. After that, the ionic gates begin their work, which are capable of moving, stopping and injecting the ionic clot, depending on the given algorithm. Further, the ion current amplifier is turned on and the collection of spectrometric data occurs (spectrometry). The recording of spectrometric data requires accurate and stable timing for 50-70 ms. After the process of ion mobility spectrometry, the amplifier is turned off and another high voltage level calibration is performed. Then the high voltage polarity is switched for a duration of 10 ms and the spectrometric cycle starts anew. Thus, the basic requirement for a synchronization system is the exact observance of time intervals.

The main requirements of additional electronics are the compatibility of the parameters when operating with the current system, the flexible tunability of the units, the ability to perform diagnostics, as well as self-testing and protection of electronic units.

Since in the classical system each controller was responsible for controlling each unit with its own unique address, then to implement compatibility with such a system it is necessary to leave the logical level unchanged, that is, to apply virtualization of controllers (Figure 2). More powerful hardware resources of the current system allow to combine several controllers into a single more efficient one.

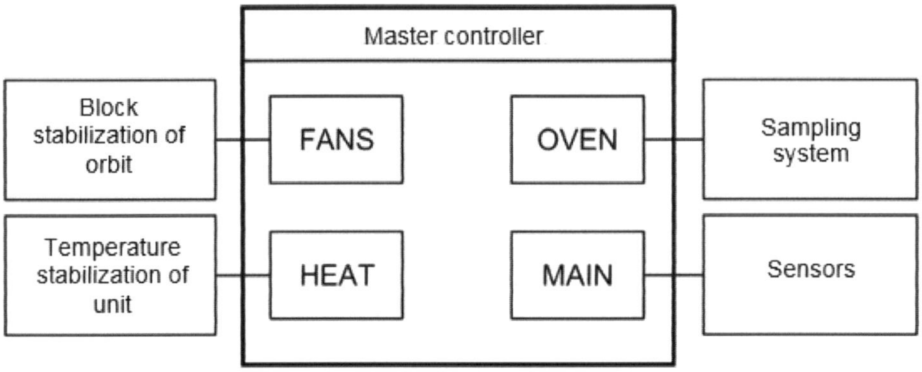

Figure 2. Block diagram of the structure of the master controller.

3. Generation of control signals for the collection of spectrometric data, ion source and gates

The structure of the spectrometric data acquisition unit is shown in Figure 3. To integrate the received data, an accurate countdown is required to avoid distortion of the spectrogram along the horizontal time axis. Therefore, for timing the system of the digitization of spectrometric data is used a timer (TIMER).

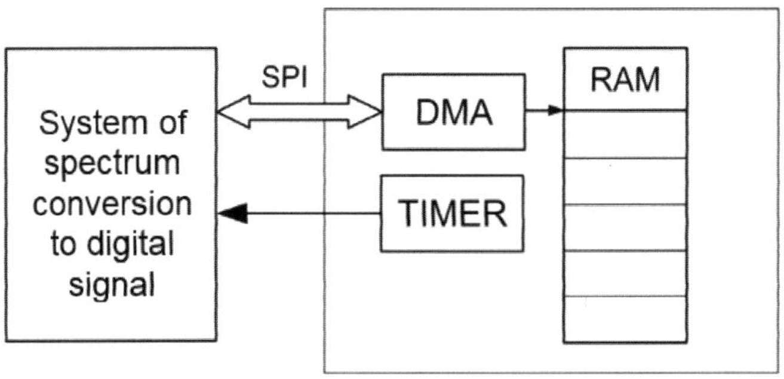

Figure 3. The unit for collecting of spectrometric data.

The IMS data is received via the SPI interface using the Direct Memory Access (DMA) controller. This allows for the increase of the performance of the controller by means of using hardware. The construction of the spectrometric data acquisition unit is shown in Figure 4.

Figure 4. Design of the collection unit of spectrometric data.

To improve the detection parameters, it is necessary to create certain conditions in the ionization chamber so as to maintain the set temperature, humidity, pressure, among other parameters. Therefore, a number of sensors are installed inside the device, which must be cycled. The device makes use of digital sensors of temperature, pressure, humidity and other values with an I2C interface. This interface is used in order to save hardware because only one microcontroller is necessary for the control of a whole set of devices and the serial data bus for communication uses only two bidirectional communication lines (SDA and SCL). Additionally, the standard provides for "hot" connection and disconnection of devices during the operation of the system. The structure of the unit for sensor questioning is shown in Figure 5.

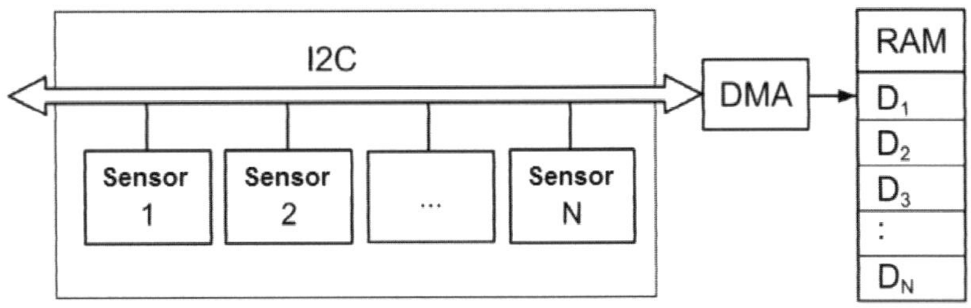

Figure 5. The structure of the unit for sensor questioning.

The controller of the direct memory access (DMA) allows to transmit information that is received from the sensors, bypassing the processor. The use of this mechanism can significantly heighten the transmitting capability and increase productivity. For this is sufficient to take the initial measure: to determine the type of transmission, provide the initial address and size of the data massive.

As in the system are used sensors of differing types, it is necessary to work out an algorithm of the data gathering and a system of the parameters for the microcontroller. The set of the parameters is given in Table 1.

Table 1. Setting parameters for the information gathering of sensors.

Sensor number	I2C Address	Qty. of bytes	Address of memory	Type
1	0b10011000	2	(u32)&Temperature3	1
2	0b10010100	2	(u32)&Temperature4	1
3	0b10011100	2	(u32)&Heater1_Temperature	1
4	0b10011010	2	(u32)&Heater2_Temperature	1
5	0b01010110	4	(u32)&Temperature1	2
6	0b01010110	4	(u32)&Humidity1	3
7	0b01010010	4	(u32)&Temperature2	2
8	0b01010010	4	(u32)&Humidity2	3
...	...	...	...	...

The given table of parameters is loaded into the microcontroller. The algorithm consists of the following sequential actions:

- Controller for the direct access to the memory counts the given quantity of bytes by the I2C address.
- The sensor type determines the formula by which are worked out the gathered data.
- Further, the information is recorded by the indicated address of the operating memory.

The electronic subsystem, coming into the components of the equipment, contains a multichannel temperature stabilization unit, which fulfills the following functions:

- Providing of the temperature mode of the sampling device.
- Providing of the temperature mode of the ionic source and gas channel of the sample input.
- Providing of the temperature mode.

Multichannel of the stabilization of temperatures allows providing for the even heating of the various zones of the. In the system are used a total of 9 temperature stabilization channels. For each channel are used separate settings of maximal power (maximal – of the controlling signal, entering the, fulfilled on the basis of the field transistor) and required temperature.

The control devices are divided into three units: heaters of the drift region, heaters of the, pumps. Each unit contains within itself the required quantity of devices and settings.

Air pumps are used for the formation of gas currents in the drift area and during the sample input. The unit for the regulation and stabilization of rotations has the following settings: work mode, starting PWM and the required speed of rotation (rot/min).

4. Monitoring of the parameters of the functioning of the ion mobility spectrometer and protection of the electron units

For the precise detection of substances and elimination of the chance of false alarms is necessary the constant control of an array of parameters and cyclical questioning of the ion mobility spectrometers.

Parameters of the equipment can be divided into several classes:

1. parameters used in ion mobility spectrometry (pressure, temperature, et al);
2. parameters determining the workability of the equipment (humidity, speed of gas currents, et al);
3. parameters of protection.

Monitoring of the atmospheric pressure of the surrounding air allows to provide correcting coefficients in the data processing program, allowing to detect substances by the data introduced into the database outside of the dependence from weather-climactic conditions.

For the channel of sample transfer and drift area the optimal temperature is 100 °C and more. This is provided for by the fact that under such temperatures the main substances under analysis weakly absorb onto inner surfaces, which provides for small losses, fast cleaning and good dynamic properties during the detection of the probe.

The charge of an ionic bunch can be estimated when the area under the spectrum of the ion current spectrum is calculated. This value affects the signal-to-noise ratio, and hence, the detection accuracy.

The device periodically conducts a self-test to check the operability and protection of electronic components. Most parameters have a range of normal operation. If you exceed the range and reach a critical value, you must take the appropriate action. These parameters are given in Table 2.

Table 2. The ion mobility spectrometer protection parameters.

Parameter number	Parameter name	Nominal value	Critical value
1	Consumption current	8 A	14 A
2	Current consumption of high-voltage unit	70 mA	200 mA
3	Current consumption of heaters per channel	1.5 A	2.5 A
4	Value of high voltage	2.2 kV	2.6 kV
5	Voltages on the collector grid	85 V	100 V
6	Temperature of the sampling device	200 °C	230 °C
7	Temperature of heaters	100 °C	115 °C

In dependence of the rank of the event, the executive controller can undertake the following actions:

- signalize to the main program about the mistake;
- block the channel;
- disconnect the device;
- completely turn off the equipment.

5. Conclusion

The article describes control system of bipolar high-resolution ion mobility spectrometer for the express detection of ultra-small quantities of substances. System includes universal microcontroller, multi-threaded control system for electronic nodes of the ion mobility spectrometer with fast polarity switching. The paper presents the algorithms for the operation and interaction of the electronic components of a bipolar ion mobility spectrometer, allowing to increase the detection efficiency of substances.

6. References

[1] Eiceman G A and Karpas Z 2014 *Ion Mobility Spectrometry (Second edition)* (CRC Press) 26
[2] Sielemann S, Li F, Schmidt H and Baumbach J I 2001 *IJIMS* **4(2)** 81
[3] Sysoev A A, Poteshin S S, Chernyshev D M et al 2014 *Eur. J. Mass Spectrom.* **20(2)** 185
[4] Podlepetsky B and Samotaev N 2016 *Lecture Notes of the Institute for Computer Sciences, Social-Informatics and Telecommunications Engineering* **170** 217
[5] Sferopoulos R 2008 *A review of chemical warfare agent (CWA) detector technologies and commercial-off-the-shelf items* (Australia: Defence Science and Technology Organisation DSTO-GD-0570) 28
[6] Ehlert S 2014 *Detection of Security-Relevant Substances via Soft Ionization Mass Spectrometric Methods Using Ambient Pressure Laser Desorption Cumulative Dissertation for the academic degree of a doctor rerum naturalium of Mathematics and Natural Sciences* (Rostock: the University of Rostock, Rostock)
[7] Jennings W, Mittlefehldt E and Stremple P 1997 *Analytical Gas Chromatography (Second Edition)* (Academic Press) 55
[8] Kenndler E 2004 *Gas Chromatography* (Vienna: Institute for Analytical Chemistry, University of Vienna)
[9] Van Bramer S E 1998 *An Introduction to Mass Spectrometry* (Chester: Widener University, Department of Chemistry)
[10] Browne C, Stafford K and Fordham R 2006 *Irish Veterinary J.* **59(2)** 97
[11] Gazit I and Terkel J 2003 *Appl. Animal Behaviour Sci.* **81** 149
[12] Gazit I and Terkel J 2003 *Appl. Animal Behaviour Sci.* **82** 65
[13] Oxley J C and Waggoner L P 2009 *Aspects of Explosives Detection* 27
[14] Diederich C and Giffroy Jean-Marie 2006 *Appl. Animal Behaviour Sci.* **97** 51
[15] O'Connor M B, O'Connor C and Walsh C H 2008 *Irish Veterinary J.* **177** 155
[16] McGann W J, Perry K J and Lepine Sh L 2008 *Trapping materials for trace detection systems* (Patent № US 7338638 B2)
[17] Linker K L, Hunter J A and Brusseau Ch A 2013 *Screening portal, system and method of using same* (Patent № US 8429987 B1)
[18] Jones D A, Gresham Ch A, Basiliere M L, Spates J J and Philip J. Rodacy 2014 *Screening system and method of using same* (Patent № US 8695443 B1)

Gallium Nitride FET Model

V V Orlov[1] and G I Zebrev[1]

[1]National Research Nuclear University MEPhI (Moscow Engineering Physics Institute), Kashirskoe shosse 31, 115409 Moscow, Russian Federation

Corresponding author's e-mail address: gizebrev@mephi.ru

Abstract. We have presented an analytical physics-based compact model of GaN power FET, which can accurately describe the I-V characteristics in all operation modes. The model considers the source-drain resistance, different interface trap densities and self-heating effects.

1. Introduction

Gallium nitride (GaN) high electron mobility transistor (HEMT) technology has many advantages, that make it a promising candidate for high-speed power electronics. It allows high-power operation at much higher frequencies than silicon laterally diffused metal-oxide-semiconductor field-effect transistors (LDMOSFETs), currently a staple for the cellular base station industry [1]. The high breakdown voltage capability (over 100 V), high electron mobility, and high-temperature performance of GaN HEMTs are the main factors for its use in power electronics applications. Circuits design in both application regimes requires the accurate compact device models that can describe the non-linear I-V characteristics. The current state-of-the-art GaN power transistor circuit models are mostly empirical in nature and contain a large number of fitting parameters. The source-drain series resistance and self-heating make the compact modeling difficult [2]. Currently available models are not enough accurate to describe the I-V characteristics of power GaN HEMTs in all operation modes. This means, that we need a compact physics-based analytical model based on the physical description of the device. In this paper, we present a physics-based GaN power transistor model based on generic approach. The paper contains 3 parts. In the first part, we will give a concise description of the model. The specific power HEMT's effects, such as series resistance and self-heating will be discussed in the second and third parts.

2. Compact model Gallium Nitride FET

2.1. Concise description of the compact model

Presented in this paper model relies on the diffusion-drift transistor model, originally proposed in [3,4]. This model is based on an explicit solution of the channel current continuity equation, and it has been consistently realized for different types field-effect devices including SOI and double-gate transistors [5], graphene FETs [6], and molybdenite MoS_2 monolayer transistor [7]. This model is able to describe via a single analytic expression the whole I-V characteristics both in subthreshold and above threshold regions and in linear and saturation modes.

A generic form I-V characteristics of an ideal (i.e., without any geometrical short-channel effects) FET in a concise form:

$$I_D = I_{DSAT}\left(1-\exp\left(-2\frac{V_{DS}}{V_{DSAT}}\right)\right). \tag{1}$$

where I_{DSAT} is the saturation current, V_{DS} is the drain-source bias. The drain saturation voltage V_{DSAT} is an explicit function of the mobile charge density:

$$V_{DSAT} \cong \varphi_T\left(1+\frac{C_{it}}{C_{ox}+C_D}\right)+\frac{qn_{S0}}{C_{ox}+C_D}, \tag{2}$$

where q is the electron charge, n_{S0} is the channel carrier density near the source, $\varphi_T = k_B T/q$ is the thermal potential, C_{ox}, C_D are the oxide and the depletion layer capacitance per unit area, correspondingly. The MOSFET electrostatic saturation current I_{DSAT} is represented as a sum of the diffusion and the drift components [8]:

$$I_{DSAT} = \frac{W}{L}qD_0 n_{S0} + \frac{W}{L}\frac{\mu_0 q^2 n_{S0}^2}{2(C_{ox}+C_D)}, \tag{3}$$

where W and L are the channel's width and length, q is the electron charge, μ_0 and D_0 are the channel carrier mobility and diffusivity, coupled by the Einstein relation $D_0 = \mu_0\varphi_T$ .The first term in (3), corresponding to a linear dependence of the drain current on n_{S0}, is a diffusion current, dominating in the subthreshold region at low n_{S0}. The second term in (3) corresponds to the saturated drift current in the above-threshold mode in a well-known square-law approximation [9]. The dependence on the gate voltage V_G has in (1) an implicit form via the dependence $n_{S0}(V_G)$. A concrete form of $n_{S0}(V_G)$ should be determined by electrostatic consideration, which is different for different device configurations (bulk or SOI FETs, double gate FETs, FinFETs etc.):

$$I_D = \frac{W}{L}e\mu_0 n_{S0} V_{DS} \equiv G_D V_{DS}, \tag{4}$$

where G_D is the channel conductance (i.e. total inverse channel resistance at low drain bias) in a low drain bias regime:

$$G_D = \frac{W}{L}e\mu_0 n_{S0}. \tag{5}$$

2.2. Channel carrier density as function of gate voltage

We will use a phenomenological interpolation for $n_{S0}(V_G)$ which is similar to that used in CMOS design compact model BSIM [10]:

$$en_{S0} = \frac{2C_1 S\ln\left[1+\exp\left(\frac{V_G-V_T}{2S}\right)\right]}{1+2\frac{C_1\,S}{C_2\varphi_T}\exp\left(-\frac{V_G-V_T}{2S}\right)}, \tag{6}$$

where V_T is the threshold voltage, C_1 (C_2) is the effective gate-to-channel capacitance per unit area in the above threshold (subthreshold) operation modes, $S = dV_G/d(\ln I_D)$ is the logarithmic subthreshold slope [9] which to be fitted from the sub-threshold part of the I-V curve. The interpolation (6) is validated by its asymptotic in the subthreshold and above-threshold regions (note $SS \equiv \ln 10 S$):

$$en_{S0} \cong \begin{cases} C_1(V_G - V_T), & V_G > V_T, \\ C_2\,\varphi_T \exp\left(\dfrac{V_G - V_T}{S}\right) = C_2\,\varphi_T 10^{\frac{V_G - V_T}{SS}}, & V_G < V_T. \end{cases} \quad (7)$$

In GaN FET the subthreshold slope is strongly dependent on the drain voltage:

$$S = S_0 + m_d V_{DS}, \quad (8)$$

where S_0 (~ 90-120 mV/decade) is the constant subthreshold slope extracted at small V_{DS}, m_d (~0.01-0.2) is a fitting constant. The threshold voltage is dependent on V_{DS} via the so-called DIBL (Drain-Induced Barrier Lowering) effect:

$$V_T = V_{T0} - \alpha_{DIBL} V_{DS}, \quad (9)$$

where V_{T0} is a constant threshold voltage, α_{DIBL} is a DIBL sensitivity parameter.

3. Accounting for series resistance

Most of the state-of-the-art power FETs are not self-aligned and their access regions affect device performance. The external drain voltage V_D can be written as a sum of internal channel voltage V_D^{int} and the voltage on R_D:

$$V_D^{\text{int}} = V_D - I_D R_D, \quad (10)$$

where R_D is the drain resistance. Taking into account the equation (1), one can rewrite (10) as follows:

$$\frac{V_D - V_D^{\text{int}}}{R_D} = I_{DSAT}(V_G)\left(1 - \exp\left(-2\frac{V_{DS}^{\text{int}}}{V_{DSAT}(V_G)}\right)\right), \quad (11)$$

where:

$$I_{DSAT} = \frac{1}{2} G_D V_{DSAT}. \quad (13)$$

From the equation (11) we have the exact solution for the drain current:

$$I_D \cong I_{DSAT}\left(1 - \frac{1}{s} W\left[s \exp\left(s - \frac{2V_{DS}}{V_{DSAT}}\right)\right]\right), \quad (14)$$

where $s = G_D R_D$, $W(s)$ is the Lambert function, defined as a solution of the equation:

$$W(se^s) = s. \quad (15)$$

IOP Conf. Series: Materials Science and Engineering **475** (2019) 012007 doi:10.1088/1757-899X/475/1/012007

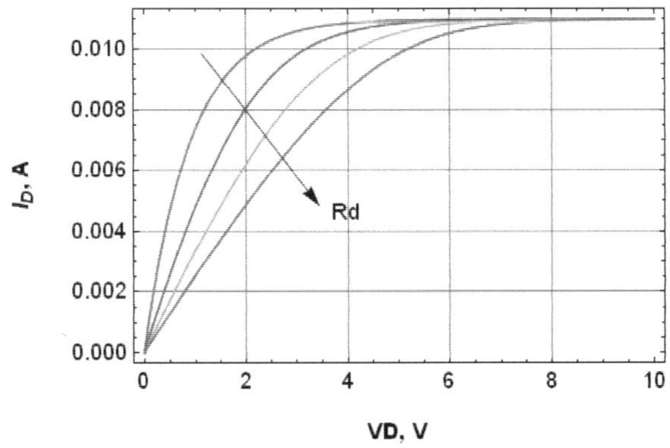

Figure 1. The simulated at $V_G = 3$ V output I-V characteristics for power GaN FET for drain resistances R_D from 0 to 300 Ohm with the step 100 Ohm. The model parameters: $W/L = 100$, $t_{ox} = 20$ nm, $\mu_0 = 300$ cm^2/(V·s), $V_T = 0.7$ V, $T_0 = 300$ K.

Figure 1 shows the output current-voltage characteristics for power GaN FET. Note, that the drain current decreases with the increasing of drain resistance R_D.

4. Self-heating effects

The problem of self-heating in power GaN HEMTs is a severe issue that that affects the device reliability and degrades the I-V characteristics [11]. The carrier mobility at room and elevated temperatures T is mainly determined by phonons, and it can be simulated as to be proportional to T^{-n} where n is a parameter (~1.0-1.5). Mobility is generally reducing the temperature due to phonon scattering:

$$\mu_0(T) \propto \left(\frac{T_0}{T_0 + \Delta T}\right)^n \propto 1 - n\frac{\Delta T}{T_0}, \tag{16}$$

where T_0 is a nominal operating temperature. We will model the self-heating effects through the following self-consistent equation:

$$I_D(T_0 + \Delta T)V_{DS}R_{th} = \Delta T, \tag{17}$$

where R_{th} is the thermal resistance (~ 30 K/W). From the equation (17) one gets:

$$I_D(T_0)\left(1 - n\frac{\Delta T}{T_0}\right) \cong \frac{\Delta T}{V_{DS}R_{th}}, \tag{18}$$

$$\Delta T \cong I_D(T_0)\frac{V_D R_{th}}{nI_D(T_0)V_{DS}R_{th} + 1}. \tag{19}$$

Here, n can be considered as a fitting dimensionless parameter of order unity. Substituting (19) to the equation (17), the drain current with self-heating effects $\tilde{I}_D$ can be approximated by the following relationship:

$$\tilde{I}_D = I_D(T_0 + \Delta T) \cong I_D(T_0) - n\frac{I_D(T_0)}{T_0}\Delta T = \frac{I_D(T_0)}{1 + n\, I_D(T_0)V_{DS}R_{th}/T_0}. \tag{20}$$

Figure 2 shows the dependencies of the drain currents I_D as functions of drain voltages V_D at different carrier mobility with the self-heating effect. The drain current decreases with the increasing of drain voltage in the saturation mode.

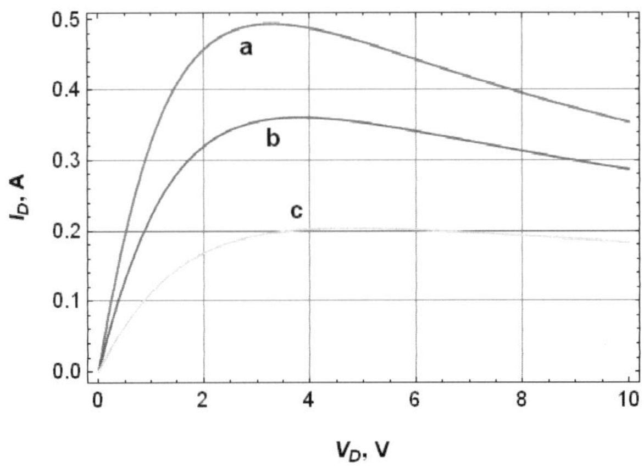

Figure 2. The simulated output I-V characteristics at $V_G = 5$ V for different carrier mobility: μ_0 = 3000 cm^2/(V·s) (a), 2000 cm^2/(V·s) (b), 1000 cm^2/(V·s) (c). The model parameters: W/L = 100, t_{ox} = 20 nm, V_T = 0.7 V, T_0 = 300 K, R_{th} = 30 K/W, n = 1.5.

5. Conclusions

Our goal has been to provide a description of the power GaN HEMT compact analytical model, which is capable of simulating in a unified manner all the FET's operation modes using a compact and transparent analytical relationship. It has been shown that the model is suitable for accurate simulation of the GaN I-V characteristics in modern power HEMT circuits. The simulation results show that the analytical model is useful for describing the I-V characteristics, taking into account the drain resistance and self-heating effects.

5. References

[1] Dunlevy L, Baylis C and Curtice W 2010 *IEEE Microwave Magazine* **11** 82
[2] Gildenblat G 2010 *Compact modeling: Principles, Techniques and Applications* (*Springer Science)*
[3] Zebrev G I, Useinov R G 1990 *Sov. Phys. Semiconductors* **24(5)** 491
[4] Zebrev G I 1992 *Sov. Phys. Semiconductors* **26(1)** 47
[5] Zebrev G I et al 2006 *SPIE proc.* **6260** 62601
[6] Zebrev G I et al 2013 *IEEE Trans. Electron. Dev.* **60(6)** 1799
[7] Sheredeko G S, Zemtsov K S and Zebrev G I 2014 *MIEL Proceed* 107
[8] Zebrev G I 2014 *SPIE Proc.* **9440** 94401
[9] Sze S M, Ng K K 2007 *Physics of Semiconductor Devices (Wiley-Interscience)*
[10] Zebrev G I, Orlov V V, Bakerenkov A S, Felitsyn V A 2017 *IEEE Trans. Nucl. Sci.* **64(8)** 2212
[11] Bagnall K R 2013 *PhD thesis Massachusetts Institute of Technology*

Research of influence of laser annealing parameters on structural and morphological properties of TiO_2 thin films

S P Malyukov[1], A V Sayenko[1] and Yu V Klunnikova[1]

[1]Southern Federal University, Institute of Nanotechnologies, Electronics and Equipment Engineering, Department of Electronic Apparatuses Design, Taganrog 347928, Russia

Corresponding author's e-mail address: sayenkoav@gmail.com

Abstract. The TiO_2 thin films are widely used as a transparent layer of n-type conductivity in the perovskite solar cells. Nanocrystalline TiO_2 films were deposited on fluorine-doped tin oxide (FTO) coated glass substrates by centrifugation and subsequent laser annealing with a wavelength of 1064 nm. The influence of laser annealing on the grain size in the TiO_2 film and spinning rate on the thickness of TiO_2 film was investigated. It was found that the grain size in the TiO_2 film ranges from 17 nm to 64 nm by using the laser annealing power of 30-70 W. The thickness of TiO_2 film changes in range from 72 nm to 124 nm depending on the spinning rate. The optimum parameters of TiO_2 thin film formed by laser annealing can increase the perovskite solar cell efficiency.

1. Introduction

Recently, nanocrystalline titanium dioxide thin films (TiO_2) have been intensively investigated, due to the perspective use of this material in the field of solar energy, photocatalysis, etc. The compact (non-porous) TiO_2 films are widely used as a transparent material of n-type conductivity and hole blocking layer in the perovskite solar cells. These cells are the most researched and promising because their energy conversion efficiency has rapidly increased from 3.8% to 19.3% within the last 5 years. The TiO_2 films' formation with a low number of pores (or cracks) of the nanometer size is important for perovskite solar cells. It allows to decrease the recombination losses and increase the energy conversion efficiency of solar cells. In general, the thickness of a compact TiO_2 film formed by spin-coating is within 50 to 150 nm [1, 2].

The method of compact TiO_2 thin film formation for application in the perovskite solar cells is described in this work. It includes spin-coating of a TiO_2 precursor and subsequent laser annealing with a wavelength of 1064 nm. The research of laser annealing power and the influence of spinning rate on structural and morphological properties of TiO_2 thin films (grain size in the TiO_2 film and its thickness) were conducted.

Laser annealing is a promising technology that finds more and more applications in micro- and nanoelectronics. Laser annealing can improve the degree of crystallinity, lower surface roughness, enlarge grain size, increase uniformity, reduce recombination centers, and also increase compactness (continuity) [3]. Thus, effectively selected parameters of laser radiation can improve solar cell performance.

2. Experiment

The uniform compact TiO_2 films were deposited on fluorine-doped tin oxide (FTO, Sigma-Aldrich) coated glass substrates by spin-coating (SpinNXG-P1, Apex) of 0.15 M titanium diisopropoxide bis (acetylacetonate) (Sigma-Aldrich) solution in ethanol at 3000 rpm for 30 s. After this, the films were heated in oven at 120°C for 5 min and then annealed by laser for crystallization [4-7].

Laser annealing was carried out with the LIMO 100-532/1064 system. It consists of a high-power infrared laser with a wavelength of 1064 nm (Nd:YAG laser, 110 W), a two-coordinate scanner, a heating element and a control computer. In this process, the laser beam was moved along the TiO_2 film surface with the help of the scanner (XY galvanometers). The film was annealed by laser radiation at the power of 30, 50 and 70 W for a period of 60-90 s. It corresponds to the surface temperature of 400-540°C [6, 7]. The crystalline phase of TiO_2 anatase is more preferable than the rutile one. The electron transport is faster in the anatase phase. However, the anatase phase irreversibly transforms into the rutile phase at temperatures above 500 °C [2, 8]. It is necessary to note that a glass substrate with TiO_2 film was preheated to 250-300°C in order to prevent thermal shock during the laser annealing.

For the research of the influence of spinning rate (2000-5000 rpm for 30 seconds) on the thickness of nanocrystalline TiO_2 films, a laser annealing process was carried out at radiation power of 50 W, which corresponds to the surface temperature of the TiO_2 film of about 470°C [6, 7].

The surface morphology and thickness of the TiO_2 film was studied by scanning electron microscopy (SEM) using a Nova Nanolab 600 (FEI Company) microscope. The optical transmission spectra of the TiO_2 film were determined by a SF-26 spectrophotometer in the wavelength range of 200-1100 nm.

3. Results and discussion

Figure 1 shows the surface morphology of TiO_2 films at varying laser annealing power (30, 50 and 70 W).

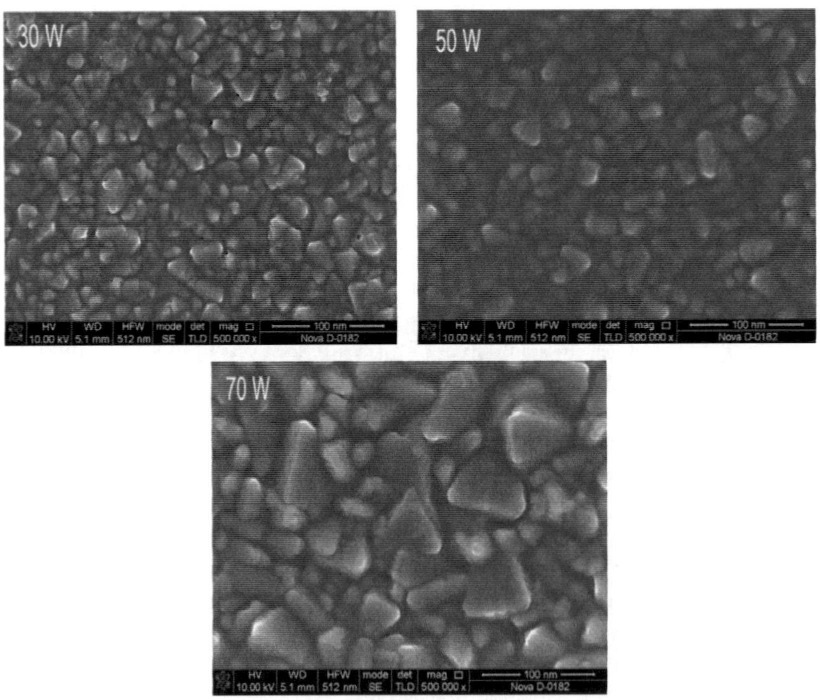

Figure 1. SEM images of the surface morphology of TiO_2 films.

It can be seen that increasing laser annealing power leads to an enlargement of TiO_2 grain size. At the laser power of 30 W, the average grain size was 17 nm. At 50 W, the grain size was 23 nm. At 70 W, it was about 64 nm. We can observe the significant growth of grain size and increase of the relief (due to coalescence [2, 9]) at the laser power of 70 W.

Figure 2 shows SEM images of cross-sectional TiO_2 films at varying spinning rates (2000, 3000 and 5000 rpm) and the laser annealing power of 50 W. It was found that the thickness of TiO_2 compact film ranges from 72 nm to 124 nm, depending on the spinning rate.

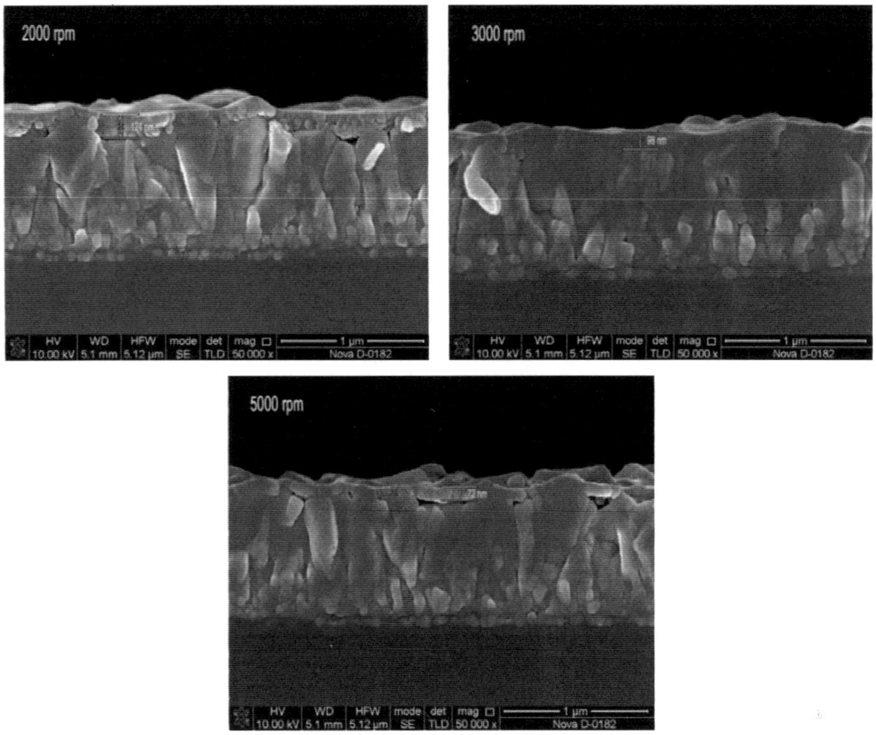

Figure 2. SEM images of cross-sectional TiO_2 films on FTO/glass substrates.

The thickness of compact TiO_2 film has a significant effect on the solar cell performance, in particular, on its energy conversion efficiency [1, 10]. The thicker the TiO_2 film (more than 100 nm), the higher the resistance to charge transport, because electrons must pass a larger distance from the perovskite film to FTO. Also, a thicker compact TiO_2 film reduces the light absorption by the perovskite film. In turn, a decrease of the TiO_2 film thickness (less than 50 nm) can lead to the incomplete coating of FTO, which can result in the undesired contact of the perovskite film with FTO and also a significant increase of charge carrier recombination. Therefore, the compact (non-porous) TiO_2 film with optimized thickness allows to achieve a higher energy conversion efficiency.

The studies of SEM images (Figure 2) showed that some pores can be formed in the TiO_2 thin films (thickness less than 80 nm), whereas cracks on the surface can be formed in thicker films (thickness greater than 120 nm). Thus, the optimum thickness of compact TiO_2 film is near 100 nm, at which conditions the film surface is the most uniform.

Figure 3 shows the transmission spectra of the TiO_2 film deposited on FTO-coating glass substrate by spin-coating at 3000 rpm and laser annealing power of 50 W. The average transmittance of about 83% in the visible and near-infrared region with a sharp absorption edge at 350 nm was observed. The excellent steepness of the optical absorption edge gives clear evidence of the high quality of compact

TiO_2 film. Moreover, the relatively high transmission of TiO_2 film indicates a low surface roughness and good homogeneity.

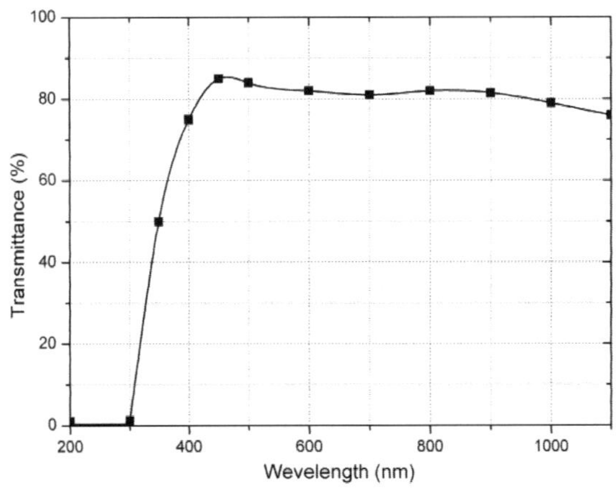

Figure 3. Transmission spectra of the TiO_2 film on FTO/glass substrate.

4. Conclusion

The method of compact TiO_2 thin film formation with the use of spin-coating and infrared laser annealing for application in the perovskite solar cells is presented. Laser annealing leads to a significant reduction of time (up to several minutes) and to a lower substrate temperature (250-300 °C) for crystallization of TiO_2 in anatase phase, in comparison with annealing in a muffle furnace (which requires up to 3 hours at temperatures of 400-500°C [2, 7])

The influence of laser annealing and spin-coating parameters on the structural and morphological properties of TiO_2 thin films was researched. It was found that the grain size in the TiO_2 film ranges from 17 nm to 64 nm at the laser annealing power of 30-70 W. The thickness of TiO_2 film changes in the range from 72 nm to 124 nm, depending on the spinning rate, which varies from 2000 to 5000 rpm. The analysis of SEM images shows that the optimum thickness of compact TiO_2 film is about 100 nm. Thus, the optimum structural and morphological properties of TiO_2 thin film using laser annealing can increase the perovskite solar cell energy conversion efficiency.

These results were obtained, using the equipment of the Research and Educational Centers «Laser Technologies» and «Nanotechnology», Institute of Nanotechnology, Electronics and Electronic Equipment Engineering of the Southern Federal University (Taganrog, Russia).

5. Acknowledgments

The study was financially supported by the Russian Foundation for Basic Research (project No 16-38-00204 mol_a).

References

[1] Jung H S, Park N-G 2015 *Small* **11(1)** 10
[2] Pugachevskiy M A 2012 *J. Tech. Phys. Lett.* **38(7)** 56
[3] Mincuzzi G, Palma A L, Di Carlo A, Brown T M 2016 *ChemElectroChem* **3** 9
[4] Avci N, Smet P F, Poelman H et al 2009 *J. Sol-Gel Sci. and Tech.* **6** 11
[5] Tavakoli M M, Gu L, Gao Y et al 2015 *Scientific Reports* **5** 1
[6] Malyukov S P, Sayenko A V, Kirichenko I A 2016 *Semiconductors* **50(9)** 1198
[7] Cha S N, Lee S, Jang J E et al 2013 *App. Phys. Lett.* **103** 1
[8] Park N-G, van de Lagemaat J, Frank A J 2000 *J. Phys. Chem. B* **104** 8989
[9] Malyukov S P, Sayenko A V 2016 *Izvestiya SFedU: Engineering sciences Russia* **4** 15
[10] Malyukov S P, Sayenko A V, Ivanova A V 2016 *IOP Conf. Series: Mater. Sci.* **151** 012033

Investigation of Influence of Sapphire Crystals Shape Factor on Its Growth by Horizontal Directed Crystallization

S P Malyukov[1], D I Cherednichenko[1], Yu V Klunnikova[1] and A V Sayenko[1]

[1]Southern Federal University, Department of Electronic Apparatuses Design, Taganrog, Nekrasovsky, 44, Russia

Corresponding author's e-mail address: sayenkoav@gmail.com

Abstract. The value of shape factor allows to determine what processes (the surface or in crystal volume) define the dynamics of sapphire crystal growth by horizontal directed crystallization method and, as a result, estimate the influence of these processes on crystal quality. We receive the Stefan problem closed solution for sapphire crystals growth by horizontal directed crystallization method. It allows to investigate the growth process and the related factors (thermal stresses on different stages of growth process), melt hydrodynamics and its influence on defects formation. 1.

1. Introduction

Having considering the kinetics of the initial stage of the crystal growth in the material liquid phase, Y.I. Frenkel pointed out that not only the thermal properties of the material occurred the significant influence on this process, but also a factor, which he called the shape factor [1]. The sapphire crystals received by horizontal crystallization method (HDC) have a characteristic configuration on the initial and the later stages of growth so the study of structure formation, impurities distribution and the defects in crystal volume during growth is the interesting problem [2].

2. Model Development

According the definition [1], the value of shape factor is equal to the ratio of crystal surface area to the object volume $\xi = \frac{F}{V}$, so the value of shape factor allows to define what processes (surface or in crystal volume) define the growth dynamics and, as a result, these influence on crystal quality.

The ratio of shape factor for sapphire received by HDC method is the following:

$$\xi(\alpha, S) = \frac{1}{h}\left(1 + \frac{(1+\alpha)h}{S \cdot \alpha}\right), \quad (1)$$

where S is the axial size of crystal; h is the crystal thickness; α is the vertex angle of crystal growth zones from seed.

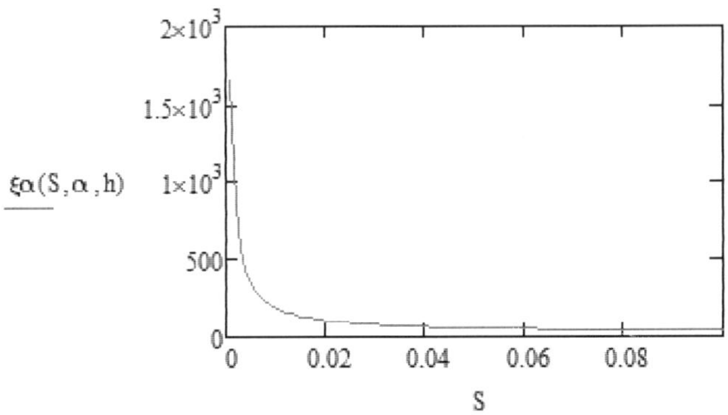

Figure 1. Dependence of crystal axial length on the form factor.

You can see that at the earlier stage of sapphire crystal growth ($S \ll h$) the processes, in particular, the latent heat release through crystal volume are dominated. At $S \gg h$ the latent heat flow (phase transformation) release is divided and is carried out through the top and the lower crystal surface. The bottom of crucible is situated under the crystal so the crucible heat conductivity and the thermal resistance of contact melt – crucible bottom will influence on growth. This fact allows to explain the incline of phase transformation surface during the process. Taking into account the shape factor it is necessary to investigate the crystal growth on all stages (growth speed, mass transfer in melt, stress, defects formation).

The shape factor indicates the level of surface crystal influence on process dynamics on all stages of sapphire crystal growth in comparison with its volume. Two main fundamental processes define the crystal growth [3]:

- law of matter conservation:

$$V = \frac{dS}{dt} = K(T_m - T_S), \tag{2}$$

- law of energy conservation:

$$V = \frac{dS}{dt} = \frac{\lambda}{L}\nabla T\big|_s - D(T_m - T_s), \tag{3}$$

where K is the kinetic parameter determining the speed of melt material transfer on surface of phase transformation; T_m is the melting temperature; T_s is the temperature of growth surface which is subject to definition; L is the latent heat of phase transformation; $\nabla T\big|_s$ is the temperature gradient in growth plane; D is the heat transfer parameter from melt on growth surface.

Proceeding from dimension of shape factor for sapphire crystal received by HDC method (1), the temperature gradient can be defined as:

$$\nabla T\big|_s = (T_s - T_0)\cdot\xi(S), \tag{4}$$

then the equation of energy conservation can be transformed to:

$$V = \frac{\lambda}{L}(T_S - T_0)\cdot\xi(S) - D(T_m - T_S). \tag{5}$$

From the energy conservation law it is possible to receive the expression for temperature of growth surface:

$$T_S = T_m - \frac{V}{K}, \tag{6}$$

and then from the equation of energy conservation it is possible to receive the ratio of growth speed as the functions of heat transfer processes and material parameters:

$$V = \frac{K(T_m - T_0)}{(1 + (\frac{\beta}{\xi}))}, \tag{7}$$

where $\beta = \frac{LK}{\lambda} + \frac{DL}{\lambda} = \frac{L}{\lambda}(K + D)$.

For simplicity it can be written as:

$$\xi = \frac{1}{h}(1 + \frac{\chi}{S}), \tag{8}$$

where $\chi = \frac{1+\alpha}{\alpha} h$.

In this case the growth speed can be written as:

$$\frac{dS}{dt} = K(T_m - T_0)\frac{(S + \chi)}{(\chi + (1 + \beta \cdot h) \cdot S)}. \tag{9}$$

We can enter the designation for convenience $\eta = 1 + \beta h$ and then the growth speed can be presented as:

$$\frac{dS}{dt} = \frac{\Delta T_0 K(\chi + S)}{(\chi + \eta S)}, \tag{10}$$

where $\Delta T_0 = T_m - T_0$ is the level of crystal initial overcooling; T_0 is the temperature of reverse side of crystal (the cold zone of growth reactor).

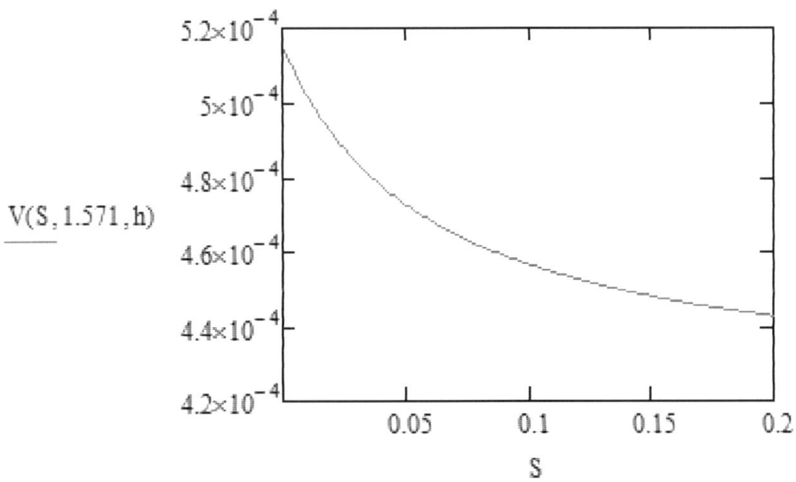

Figure 2. Dependence of growth rate on the crystal axial length.

The time for reception of crystal with length S at this level of initial overcooling can be defined by integration of formula (10):

$$t(S) = \frac{(S - S_0) - \chi \ln(\frac{\chi + S}{\chi + S_0})}{K \cdot \Delta T_0} \tag{11}$$

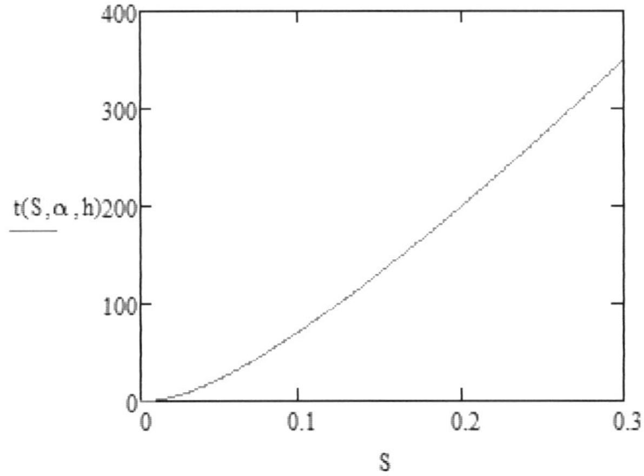

Figure 3. Dependence of growth time on the crystal axial length.

If it is difficult to determine the seed thickness S_0 (longitudinal) according the technological features of furnace so it can be neglected. The change of phase transformation surface temperature can be fined substituting the ratio (10) in formula (6).

3. The investigation results

We received the Stefan problem closed solution for sapphire crystals growth by horizontal directed crystallization method. It allows to investigate the growth process and the related factors (thermal stresses on different stages of growth process), melt hydrodynamics and its influence on defects formation and as a result on crystal quality.

Calculations are necessary for development of sapphire growth furnaces by HDC method.

The temperature of growth surface is defined by formula:

$$T_s = T_m - \frac{T_m - T_0}{1 + \dfrac{\beta}{\xi}}. \tag{12}$$

One can see that in the case of self-coordinated growth T_s changes in limites $T_s = T_0$ at $S = S_0$ (or S_0=0) and $T_s = T_m$ at S limit. The growth can stop under these conditions.

Acknowledgments

This research is supported by the Federal Target Program No. 14.587.21.0025, the project ID is RFMEFI58716X0025.

References

[1] Frenkel Y I 1975 *Kinetic Theory of Liquids* (Leningrad: Science) 592
[2] Lyubov B Y 1975 *Theory of Crystallization in Large Volumes* (Moscow: Science) 256
[3] Malyukov S P, Klunnikova Y V and Cherednichenko D I 2013 *Sapphire: Structure, Technology and Applications* (USA: Nova Science Publishers) 101

Comparative analysis of silicon nanostructures by x-ray diffraction technique

A Yu Kharin [1], R B Assilbayeva [2], Yu V Kargina[1,3] and V Yu Timoshenko [1,3]

[1]National Research Nuclear University "MEPhI", Bio-Nanophotonic Lab., 115409 Moscow, Russia
[2]Satpaev National Research Kazakh University, 050013 Almaty, Kazakhstan
[3]Lomonosov Moscow State University, Physics Department, 119991 Moscow, Russia

Corresponding author's e-mail address: AYKharin@mephi.ru

Abstract. Silicon nanoparticles and nanowires prepared by using different methods were investigated by means of x-ray diffraction (XRD) technique. Broadening of the XRD lines of crystalline Si lattice allows us to estimate the mean size of nanoparticles. Silicon nanoparticles stored in aqueous suspensions for several days were found to decrease their sizes due to dissolution process. The kinetics of dissolution for different kinds of nanoparticles were found to be different. It was also shown that the most unstable nanoparticles (microporous silicon ones) could be protected from dissolution via interaction with the polysaccharide molecules (dextrane) into the solution.

1. Introduction

Silicon (Si) nanostructures have attracted a lot of interest since the discovery of visible photoluminescence from porous silicon (PSi) [1]. PSi nanostructures have been considered for optoelectronics [2], photovoltaics [3,4], and various sensing [5,6] applications. Furthermore, it has been demonstrated a large panel of imaging and therapeutic modalities, including theranostic applications of PSi layer and Si nanoparticles (NPs) [7-9]. It was found that small Si NPs can be dissolved in water solution [10]. While the dissolution rate of bulk crystalline silicon (c-Si) is defined by the solution pH and orientation of the monocrystalline surface, the dissolution rate of NPs mainly depends on their size, morphology and surface oxide coating. Despite numerous different parameters, the most important dependence is that the small Si NPs are degraded into water solution during relatively short times (up to 24 h). The fact of fast degradation of Si NPs leads to arising from the huge number of research on the application of PSi NPs for drug delivery and controlled drug release [8]. Nevertheless, the control on the dissolution rate without changing the size of particles will be very helpful for the bio-applications of Si NPs. In the present paper, we show the possibility to monitor the mean size and dissolution rate of Si NPs by means of the XRD technique.

2. Materials and methods

Samples of microporous silicon were formed by the standard method of the electrochemical etching [1,2] of bulk c-Si wafers (p-type, boron doped to the specific resistance of 1-10 Ω*cm, (100)-oriented) in a mixture of hydrofluoric acid (48%) and ethanol (1:1). The time of etching was 40 min and the current density was 60 mA/cm^2. The free standing PSi films were obtained by lifting from the substrate after applying a short pulse of the etching current 500 mA/cm^2. Prior investigations the porous silicon films were air-dried for one week. The mesoporous silicon was obtained using the similar procedure on a heavily boron-doped p-type silicon wafer, with the specific resistivity of 1-5 mΩ*cm, (100)-oriented. Silicon nanowires (SiNWs) were obtained by using metal-assisted chemical etching of c-Si wafers of p-type, boron doped, 1-10 Ω*cm, (100)-oriented (see for details Refs. [11,12]. In order to investigate the stability of Si nanostructures in water, the corresponding powders were suspended in water. Microporous silicon (12 mg) was ground in the presence of water (1 mL) or an aqueous solution of dextrane. Powders of Si NPs prepared by microwave induced decomposition of SiH_4 [13] were also studied.

XRD spectra were measured using a RENOM apparatus (EXPERTCENTER Ltd., Russia) with Cu Kα line (λ=0.154184 nm) at room temperature in air. Measurements were done for powders just after synthesis, after mixing with water or aqueous dextrane solution followed by drying in air.

3. Results and discussion

Figure 1 shows typical XRD patterns for c-Si powder, SiNWs, meso- and micro- PSi in semi-logarithmic scale. It can be clearly seen that there is almost no difference in XRD patterns of c-Si and Si NWs within the detection limit 0.01 deg. The XRD spectrum of SiNWs consists of peaks corresponding to the presence of silver (Ag) nanoparticles used during the synthesis. The amount of silver, estimated via the corundum numbers is 2±0.5 mass %. The amount of amorphous phase corresponds to the background signal intensities, so the relative amount of amorphous oxide can be measured as a ratio between background signal and peak intensity.

Mean size of Si nanocrystals in micro- and meso-PSi can be estimated from the broadening of diffraction line by using the Scherrer equation:

$$d = \frac{K\lambda}{\Delta\theta cos\theta} \quad , \qquad (1)$$

where λ is the wavelength of x-ray radiation (λ=0.154184 nm, in our case), K is a dimensionless shape factor close to unity, $\Delta\theta$ is the line broadening at half the maximum intensity (FWHM), θ is the Bragg angle of the corresponding crystallographic plane.

According to Eq.1 the mean size of Si crystallites in micro-PSi and meso-PSi is 2.7-4.4 nm and 10-15 nm, respectively, which in accordance with the literature data for similar preparation conditions [1,2].

Figure 2 shows XRD spectra for Si NPs prepared by silane decomposition and c-Si powder (for comparison). The broadening of Si NPs lines allows us to estimate the nanocrystal mean sizes to be about 30-40 nm. The background of XRD spectrum of Si NPs indicates the presence of amorphous phases, which can be both amorphous Si and SiO_2.

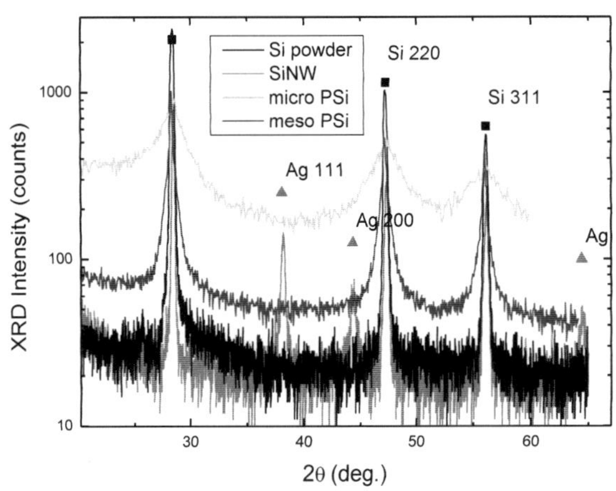

Figure 1. XRD spectra of powders of c-Si, SiNWs, meso-PSi and micro-PSi. Black squares and red triangles indicate positions of the corresponding scattering angles for c-Si and crystalline silver from the XRD database.

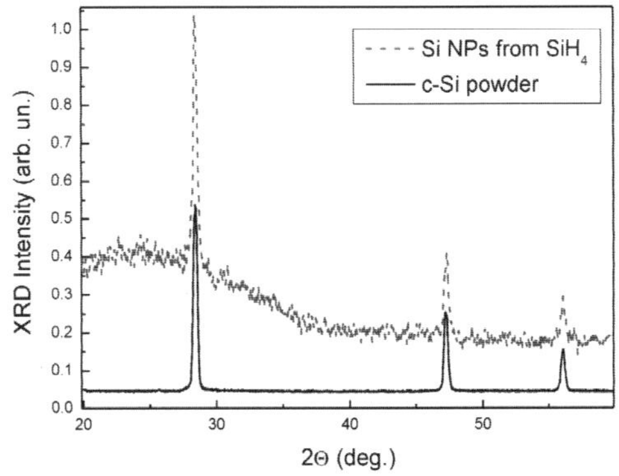

Figure 2. XRD spectra of powders of Si NPs prepared by silane decomposition and c-Si (for comparison).

According to Figure 3, the storage of Si NWs in water did not significantly change the XRD patterns, while the grinding of porous silicon in the presence of water during 10 min cause a significant change of XRD. It is known that micro-PSi can be oxidized by water [2], so the silicon peaks disappear. Instead of it, we can see an increase in the background caused by the presence of amorphous silicon oxide.

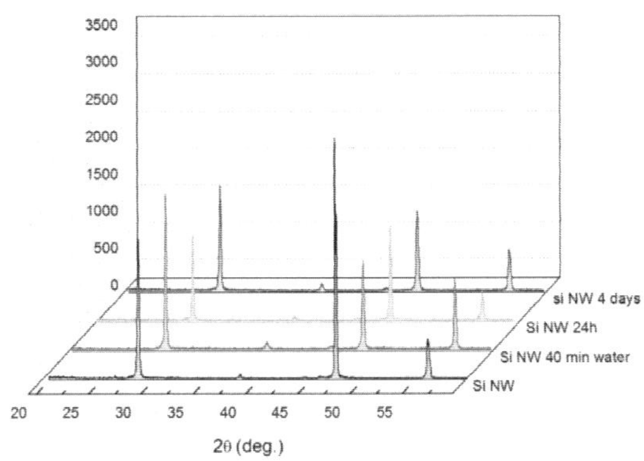

Figure 3. XRD spectra of Si NWs before and after storage in water for a different time.

Assuming that the average thickness of oxide layer at the surface of silicon at room temperature without thermal treatment usually varies from 1 to 1.5 nm and taking into account the shape and size of the particles we can calculate the ratio between Si and SiO_2 phases. It can be expressed through the bond absorption for each type of particles. The ratio of background intensity (caused mainly by the presence of amorphous SiO_2) and integral intensity of XRD silicon peak with a maximum at 28.44^O can be used to show the increasing amount of amorphous silica in nanostructures with a decrease of their size. It allows us to propose the following expression for the ratio between XRD intensities and masses of Si and SiO_2 fractions:

$$\frac{I_{background}}{I_{peak}} \sim \frac{m_{SiO2}}{m_{Si}} \quad . \tag{2}$$

Figure 4 shows the data calculated according to Eq.2 versus the mean size of Si nanocrystals.

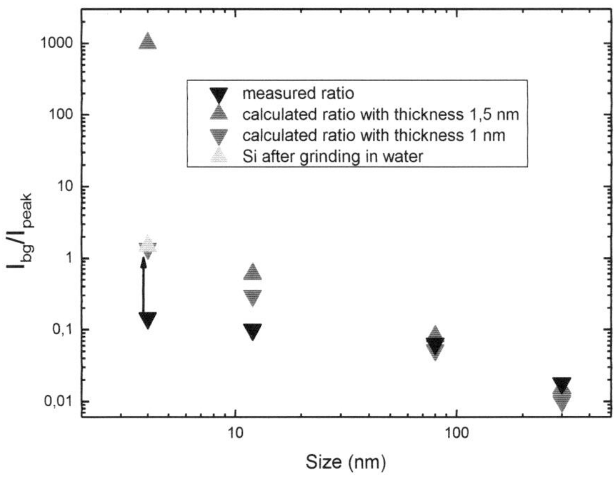

Figure 4. Dependence of the ratio between the XRD background and intensities of the characteristic c-Si lines vs. mean size of Si nanocrystals.

Figure 4 shows that the decrease of the size of Si nanocrystals results in an increase of the content of silica phase. Nevertheless, in the case of relatively big nanostructures, the oxidation layer seems to correspond the model of constant silica thickness, while in the case of porous silicon we can see that the particles are much less oxidized than they are supposed to be from the model of free-standing crystallites. That effect is probably caused by the complicated porous structure of the nanoparticles. That structure slows down the diffusion of oxygen. That effect also leads to the possibility of self-inflammation or even explosion of porous silicon during mechanical treatment. After grinding of PSi layer in water, we can observe the partial oxidation of the microporous silicon (green triangle in Fig.4) and formation of silicon oxide in amount equivalent to the formation of 1nm layer of silicon dioxide on each crystallite.

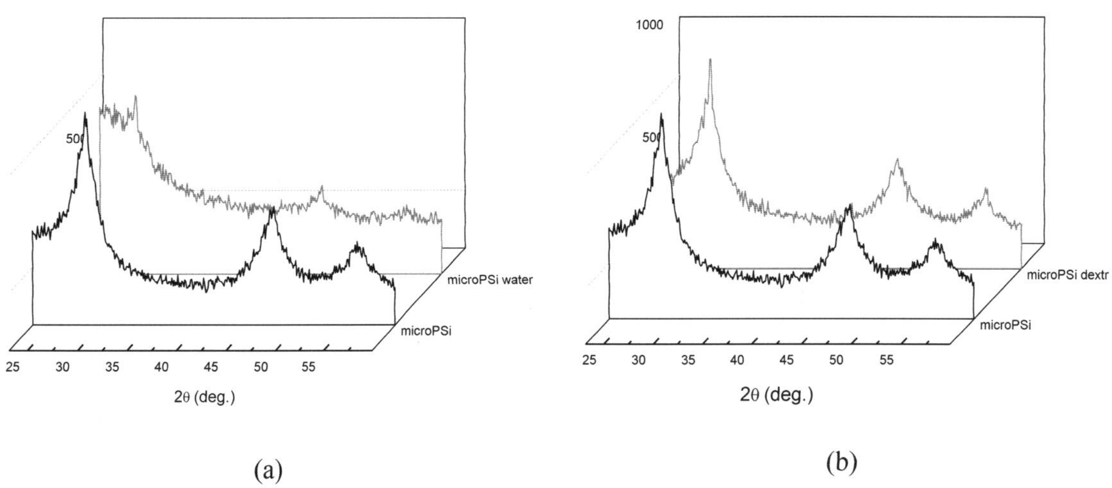

Figure 5. (a) XRD spectra of micro-PSi before and after grinding in water: (b) the same spectra for micro-PSi treated in an aqueous solution of dextrane.

One of the ways to protect Si NPs from rapid degradation in water is based on the coating of the NPs with polymer shell [13]. We used the grinding of PSi in dextrane solution in order to evaluate if the simple presence of this polymer has the protective effect. Figure 5 shows XRD spectra of porous silicon before and after grinding into the dextrane solution. It can be easily seen that there is no noticeable difference between XRD patterns. This fact is explained by the protective role of dextrane for Si nanocrystal surfaces.

4. Conclusion

We show that the XRD method, which is usually rarely used for silicon nanostructures, have great potential for the determination of the size of nanocrystals, the amount of silicon oxide (after calibration) and it can be used to monitor the stability/degradation of silicon nanostructures in aqueous solutions. The XRD measurements demonstrated a correlation between the silicon nanocrystal size and the content of the amorphous silica in different nanostructures. It was shown, that the fast oxidation of small silicon particles takes place in water, but it was shown that it could be slowed down by the dextrane coating of the particles. The obtained results can be useful for the control of biodegradation in the case of bioapplications of silicon nanoparticles.

Acknowledgments
Authors (V.Yu.T. and K.Yu.V.) acknowledge the financial support of the Russian Foundation for Basic Research (grant 16-02-00668) and R.B.A. thanks the Ph.D. Program of Satpaev National Research Kazakh University.

References

[1] Canham L T 1990 *Appl. Phys. Lett.* **57(10)** 1046
[2] Sailor M J 2012 *Porous silicon in practice: preparation, characterization and applications (Weinheim: Germany)* 254
[3] Wu C, Crouch C H, Zhao L et al 2001 *Appl.Phys. Lett.* **78(13)** 1850
[4] Halbwax M, Sarnet T, Delaporte P et al 2008 *Thin solid films* **516(20)** 6791
[5] Lin V S Y, Motesharei K, Dancil K P S et al 1997 *Science* **278(5339)** 840
[6] Konstantinova E A, Ryabchikov Y V, Osminkina L A et al 2004 *Semicond.* **38(11)** 1344
[7] Erogbogbo F, Yong K T, Roy I, Xu G, Prasad P N, Swihart M T 2008 *ACS Nano* **2(5)** 873
[8] Park J H, Gu L, Von Maltzahn G, Ruoslahti E et al 2009 *Nat. Mat.* **8(4)** 331
[9] Kabashin A V, Timoshenko V Yu 2016 *Nanomedicine* **11(17)** 2247
[10] Canham L T 2007 *Nanotechnology* **18(18)** 185704
[11] Osminkina L A, Sivakov V A, Mysov G Aet al 2014 *NRL* **9(1)** 1
[12] Timoshenko V Y, Gonchar K A, Golovan L A et al 2011 *J. Nan. and Opt.* **6(4)** 519
[13] Gongalsky M B, Kharin A Y, Osminkina L A et al 2012 *NRL* **7(1)** 1

Influence of the length of organic molecules of ligands on the PbS QD solids optical properties

Y V Goltyapin[1], M A Zvaigzne[1], A A Chistyakov[1], V R Nikitenko[1], P S Samokhvalov[1], I L Martynov[1] and A R Tameev[2]

[1]National Research Nuclear University MEPhI (Moscow Engineering Physics Institute), Moscow, 115409, Russia
[2]Frumkin Institute of Physical Chemistry and Electrochemistry of the Russian Academy of Sciences, 31-bld.4 Leninsky Prospect, Moscow 119071, Russia

Corresponding author's e-mail address: goltyav@gmail.com

Abstract. The quantum dot (QD) solids are promising materials for use in electronics. We studied the effect of variations in the length of organic ligands on the optical properties of PbS QD solids from the measurements of photoluminescence spectra and decay kinetics. The decay is becoming exponentially faster along with the decrease of the length of ligand molecules, which is consistent with the decay of excitons by hopping of charge carriers.

1. Introduction
The quantum dots has proven to be quite useful material for different applications [17, 18]. In recent years, the great attention is devoted to the optical and electrical phenomena in the so-called QD-solids. This term means a new type of a condensed matter, which is formed by "artificial atoms" – metallic or semiconducting nanocrystals. QD solids are very promising for electronics, in particular related to photovoltaics [1-4] and electroluminescence [5,6], because one can manage the optical and electronic properties due to quantum confinement effect. Charge transport is an important process for the operation of electronic devices [7,8]. Although great activity was developed in order to obtain well-ordered structures (superlattices), positional and hence energetic disorder of electronic (hole) states remain a very important factor, in particular, due to scatter in a size of QDs [7]. Another important source is a disorder in a non-ideal structure of shell around every QD, which is formed by surface-active organic molecules (ligands). These shells are considered typically as spacers, separating different QDs, hence supporting the quantum confinement, but the function of ligands in QD solids is not completely clear [9,10]. The disorder in combination with the weakness of electronic coupling of different QDs can produce a localization of all electronic (hole) states in a QD-solid [11], reminiscent to organic semiconductors [2,8]. The nature of transport in semiconductor QD solids is still under debates. It was considered as band-like process via extended states [1], while the hopping nature of transport also was supported [2,7,8]. The latter case seems to be very probable in a rather disordered QD-solids [2,7]. In this work, we replaced ligands in a PbS QD-film and analyzed the influence of variations in the length of molecules of ligands on the photoluminescence spectra and decay kinetics.

2. Experimental
In this work, QDs PbS synthesized by the colloidal chemistry method were used, and they have a typical diameter of 4 nm. The surface of QDs was covered with a monolayer of molecules of oleic acid

(OA) as a result of the synthesis. The initial ligands were replaced by new ligands, such as hydrazine, and 1,6-hexanedithiol (HDT). The lengths of new surface molecules, respectively, are 4 and 2 times smaller in comparison with the OA molecule.

Hydrazine is widely used to replace long-chain ligand molecules on the surface of PbS QDs. In turn, HDT is still poorly researched and is used for the first time, to our best knowledge, as a surface molecule for the formation of condensates of quantum dots. The presence of a thiol group in the composition of the HDT suggests that it will be used in the "role" of the ligand for QDs PbS, due to the high affinity for sulfur molecules, which allows chemically to "sew" the QDs in the film.

QD-solids with the initial ligands were thin films obtained by the spin-coating method. The replacement for new ligands was carried out directly in the process of the film preparation. A solution of QD (with OA as a ligand) in toluene with a volume of 50 μl and a concentration of 10 mg / ml was applied to a pre-cleaned glass substrate, which was then rotated for 30 seconds at an angular rate of 1500 rpm in a spin-coating setup. After that, 50 μl of the solution of the molecules of the new ligand (for hydrazine, 1 mole / L in methanol, for HDT 0.1 g / liter in acetonitrile) was applied to the resulting film using an automatic pipette, and after 30 seconds of exposure the film was rotated 30 seconds with an angular speed of 1500 rpm. Further, to remove excess molecules of ligands from the surface of the film, a suitable solvent for each ligand was applied to it, and the film was rotated for 30 seconds at a speed of 1500 rpm again. The described procedure was repeated 5 times to obtain the required thickness. The thickness of the obtained samples, 40-100 nm, was measured with a microinterferometer.

Control of replacement of surface molecules was carried out by analyzing the IR absorption spectra, obtained with the Varian IR spectrophotometer "Scimitar1000 FT-IR" for QD films.

For all samples of the films and the QDs solution, the luminescence and absorption spectra were measured, as well as the luminescence decay kinetics. For research of the absorption spectra, we used the halogen lamp Ocean Optics LS-1 for a wide spectral range, a semiconductor laser with a radiation wavelength of 908 nm was used to excite the luminescence. The spectra were recorded with the Avantes AvaSpec-NIR256-1.7 spectrometer.

For the research of the kinetics of luminescence, the second harmonic of the picosecond YAG: Nd3 + laser was used (532 nm wavelength, pulse duration 350 ps, pulse repetition frequency 50 Hz, average power 3 mW). The signal was recorded at the wavelengths of the luminescence maximums with the help of a high-speed PIN photodiode.

3. Results

As follows from Figure 1, a red shift of the luminescence spectra of the QD films was observed in the respect to the solution, which became more pronounced when the ligand was changed to a shorter one. The position of the first exciton absorption maximum remained unchanged after the transition from the solution to the condensed state. In addition, the width of the luminescence spectra varied. If the transition from a solution of QD with OA as a ligand to a film occurs, the spectrum was broadened and amounted to $\Delta\lambda = 195 \pm 8$ nm. If the original ligands were replaced with HDT in the film, the luminescence spectrum narrowed to $\Delta\lambda = 159 \pm 8$ nm, and if it was replaced by hydrazine, it significantly broadened and amounted to $\Delta\lambda = 320 \pm 8$ nm.

When measuring the kinetics of luminescence decay, the typical biexponential time dependence of decay kinetics [9] was observed in QD-films with initial ligands (OA):

$$I(t)=A_1 e^{(-\frac{t}{\tau_1})} + A_2 e^{(-\frac{t}{\tau_2})}, \quad (1)$$

where τ_1 and τ_2 represent the fast-decay and the slow-decay component, respectively; A_1 and A_2 are their amplitudes. We obtained $\tau_1 = 40$ ns and $\tau_2 = 160$ ns, while in a solution decay kinetics is monoexponential with the decay time $\tau_0 = 570$ ns, see Figure 2.

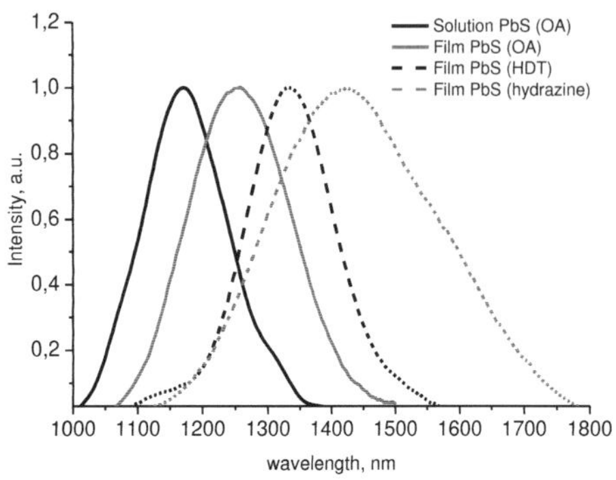

Figure 1. FL spectra of a solution and QD films with different surface ligands.

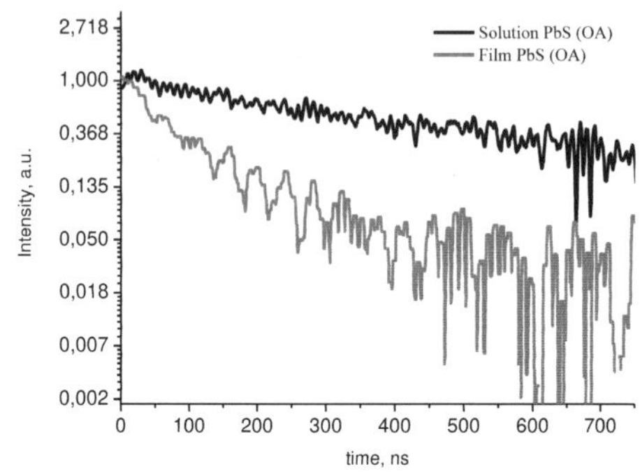

Figure 2. Kinetics of the luminescence decay of a solution and QD films with initial ligands.

A significant decrease in the decay time of luminescence was observed in the films with replaced ligands, namely 15 and 7 ns for the case of ligands HDT and hydrazine, respectively. Probably, monoexponential kinetics is observed.

4. Discussion

The width of the luminescence spectrum is directly related to the spreading of QDs over the size inside the film. The spreading could be apparently increased by an aggregation of QDs. One can assume that an aggregation is sufficient after replacing HDT ligands with hydrazine, reflected on the increased width of the spectrum, see the Figure 1, while in a QD film with a HDT, the particles are fairly uniformly distributed over the sample.

The observed redshift in the photoluminescence peaks of QDs, see the Figure 1, can be mainly due to four possible causes [12]:

a) increase in the size of QDs, that is not occurs in this case;
b) stronger electronic coupling between neighboring QDs with shorter ligands, which reduces the quantum confinement and increases the probability of charge transfer over the manifold of size-distributed QDs; an electon (a hole) is transferred most probably from a nanoparticle with a wider energy gap to a nanoparticle with a smaller energy gap at initial time interval;
c) the Förster-type resonant transfer of an exciton; this process of energy transfer will lead to the formation of an asymmetric spectrum due to the extinction of photoluminescence from the high-energy side [12]; however, the symmetry of the luminescence spectra is observed from Figure 1, precluding the exciton transfer;
d) charge transfer between QD and surface ligands; one has to note, that the charge transfer between QD and the surface ligand is not a noticeable process in the PbS QD films [13].

Therefore, one can assume that a stronger coupling of electrons and respective facilitation of charged transfer to neighbor QDs is the main factor, contributing to the red shift, observed in the luminescence of QD films with replaced ligands. This fact allows one to assume the appearance of an additional channel of nonradiative relaxation after replacing of initial ligands to the shorter ones. An analysis of the possible mechanisms of nonradiative processes under the conditions of our experiment [14,15] suggests that this is either resonant energy transfer between QDs of various sizes (excluded, as was mentioned above), or transfer of charge carriers between QDs. The transfer rate increases exponentially along with the decrease of a length of a ligand molecule, leading to the sharp increase of a photocurrent [16]. Despite the aggregation of QDs and the non-uniform spatial distribution of particles inside the film with hydrazine, the luminescence kinetics in these samples are the shortest, indicating a higher probability of the charge transfer process compared to other films. Consequently, the main factor affecting the transfer efficiency is the ligand length, which determines the distance between neighboring QDs, rather than their chemical structure or uniformity of QDs distribution within the sample.

In order to analyze evolution of characteristic times of luminescence decay along with transition from a solution to QD-film and from longer to shorter ligands in QD-films, we describe kinetics of charge carriers and excitons by the following set of equations:

$$\frac{dP}{dt} = E/\tau_d - k_m P^2 \tag{2}$$

$$\frac{dE}{dt} = -E\left(\frac{1}{\tau_d} + \frac{1}{\tau_0}\right) + k_m P^2, \tag{3}$$

where E and P are concentrations of excitons (which is proportional to luminescence intensity) and charge carriers (holes and electrons), respectively; τ_0 is a recombination time in solution (i.e. lifetime of excitons in isolated QDs), and τ_d is a transfer time of a carrier to the neighbor QD. Eq. 2 assumes that charge carriers appear due to decay of excitations, and disappear due to formation of excitons after the meeting with opposite charge, see the second term with the rate constant k_m (generation is supposed to be spatially uniform, hence transport term is absent). Eq. (3) means that decay of excitons results from production of charge pairs and recombination, the rate of the latter is the same as in solution. Biexponential character of the luminescence decay, see Eq. (1), provides that E is a solution of the differential equation $\ddot{E} + 2\beta\dot{E} + \omega^2 E = 0$, $\beta^2 - \omega^2 \geq 0$, where β and ω are some constants. Comparing this equation and Eq. (3) after differentiation in time, one require $\omega^2 E \approx -2k_m P\dot{P}$. Analysis of the system (2), (3) under this approximation yields

$$\frac{1}{\tau_1} + \frac{1}{\tau_2} = \frac{1}{\tau_d} + \frac{1}{\tau_0} \tag{4}$$

Substitution of measured values of τ_0, τ_1 , τ_2 for the case of ligands OA leads to the conclusion, that $\tau_d \approx \tau_1$ (more precisely, $\tau_d \leq \tau_1$). Eq. (4) is consistent with experimental result $\tau_0 > \tau_2$.
Since $\boldsymbol{\tau_1} \approx \boldsymbol{\tau_2}$ for the case of ligands HDT and hydrazine, and $\boldsymbol{\tau_0} \gg \boldsymbol{\tau_d}$, eq. (4) yields

$$\boldsymbol{\tau_d \approx \tau_1/2} \tag{5}$$

for these ligands. Providing hopping nature of charge separation, we assume

$$\boldsymbol{\tau_d = t_0 \exp(2\gamma l)}, \tag{6}$$

where $\boldsymbol{\gamma}$ is an inverse localization radius, l is the length of a ligand molecule, and t_0 is an attempt-to jump time. Substitution of $\boldsymbol{\tau_d}$ values from Eqs (4), (5) to Eq. (6), and $l \approx$ 2, 1 and 0.5 nm for OA, HDT and hydrazine, respectively, yields $\boldsymbol{\gamma_{opt}}^{-1} \approx \mathbf{1.2}$ nm. One has to note, that the estimation of the same value from steady-state conductivity measurements [16] results in considerably smaller value, $\boldsymbol{\gamma_{cond}}^{-1} \approx \mathbf{0.4}$ nm. , which is more consistent with the hopping nature of charge transfer. One has remind, however, that the measurements of luminescence decay shows the transient process in a disordered media on a short time scale. Charge transport should be involved in this process in an extremely non-equilibrium (dispersive) mode, while conductivity measurements [16] deals with the transport in quasi-equilibrium mode. In the first case, an average hopping frequency should be much larger, because the fastest jumps realized on a short time scale after the generation pulse. That means that the 1) hopping distances are considerably scattered (positional disorder), 2) the shortest distances are involved in hopping on a short time scale, and 3) thermally activated (hence delayed) hopping is a rate-limiting step in a quasi-equilibrium mode, due to energetic disorder. Thus, the $\boldsymbol{\gamma_{opt}}^{-1} \approx \mathbf{1.2}$ nm seems to be an apparent value, rather the shorter lengths l should be considered.

5. Conclusion

Using PbS QD solids, we demonstrated that the shorter the length of ligand molecules the faster the rate of luminescence decay. This length is considered as a measure of the mean inter-QD distance in QD solid. We propose that the reason of the correlation is the increase of an additional channel of exciton's decay, namely the exciton dissociation by the means of hopping of a charge carrier to the neighbor QD. This result agrees well with the recently reported conclusion that photocurrent increases exponentially with decreasing ligand length [16].

6. Acknowledgements

The work was supported by "improving the competitiveness" program of the National Research Nuclear University «MEPhI». Preparation of QD solids was supported by the Russian Foundation for Basic Research, grant no. 16-29-09623 (OFI_M).

References

[1] Choi J H, Fafarman A T, Oh S J et al 2012 *Nano Lett.* **12** 2631
[2] Chistyakov A A, Zvaigzne M A, Nikitenko V R et al 2017 *J. Phys. Chem. Lett.* **8** (accepted)
[3] Pathak D, Wagner T, Adhikari T, Nunzi J M 2015 *Synth. Met.* **200** 102
[4] Kramer I J, Sargent E H 2014 Chem. Rev. **114** 863
[5] Dayneko S, Lypenko D, Linkov P et al 2016 *Materials Today: Proceedings* **3** 211
[6] Dayneko S, Lypenko D, Linkov P et al 2014 *Proc. SPIE* **9270** 927009
[7] Guyot-Sionnest P 2012 *J. Phys. Chem. Lett.* **3** 1169
[8] Gilmore R H, Lee E M Y, Weidman M C, Willard A P, Tisdale W A 2017 *Nano Letters* **17** 893
[9] Dayneko S, Linkov P, Martynov I et al 2016 *Physica-E: Low-dimensional Systems and Nanostructures* **79** 206
[10] Dayneko S, Tameev A, Tedoradze M et al 2013 *Appl. Phys. Lett.* **103** 063302
[11] Yang J, Wise F W 2015 *J. Phys. Chem. C* **119** 3338
[12] Xu F, Gerlein L, Ma X, Haughn C, Doty M and Cloutier S 2015 *Materials* **8** 1858
[13] Choi J J, Luria J, Hyun B R et al 2010 *Nano Letters* **10** 1805
[14] Luther J M, Law M, Qing Song, Craig L et al 2008 *ACS Nano* **2** 271
[15] Zarghami M H, Liu Y, Gibbs M et al 2010 *ACS Nano* **4** 2475
[16] Zvaigzne M A, Alexandrov A E, Samokhvalov P S et al 2017 *Tech. Phys. Lett.* **44** (accepted)
[17] Nabiev I 2017 *J. of Phys.: Conf. Ser.* **784(1)** 25
[18] Bilan R S, Krivenkov V A, Berestovoy M A et al 2017 *ChemPhysChem* **8(18)** 970

IOP Conf. Series: Materials Science and Engineering **475** (2019) 012013 doi:10.1088/1757-899X/475/1/012013

Impact of Extrinsic Interface Traps and Doping Atoms on Conductivity of Graphene Field Effect Devices

G I Zebrev[1] and S A Shostachenko[1]

[1]National Research Nuclear University MEPHI, Moscow, Russia

Corresponding author's e-mail address: gizebrev@mephi.ru

Abstract. Near-interfacial oxide traps and chemical impurities on the graphene surface or at the graphene-dielectric interface can be a source of intentional or unintentional doping of graphene sheet. The efficiency of such chemical doping can vary in a wide range depending on parameters of graphene field effect devices. Mechanisms of such sensitivity of doping efficiency to the device characteristics need to be understood. The objective of this paper is to theoretically derive the analytical relations, adapted to the explicit calculation of graphene chemical doping.

1. Introduction

Graphene is a two-dimensional monolayer of carbon atoms arranged in a honeycomb hexagonal lattice [1]. Due to the unique structure of the lattice, graphene has a number of remarkable properties (zero bandgap, high carrier mobility etc.) which make it a promising candidate for advanced application in electronics. Despite significant progress in technology, the functional characteristics of graphene field effect device are still far from perfection. As a rule, graphene is not intrinsic material. Charged impurities in the dielectrics, or, chemical atoms or molecules on the surface of graphene sheet are able to dope the graphene, causing the charge neutrality voltage shifts and field-effect mobility and transconductance degradation. At the present, it is becoming increasingly clear that in order to fully describe the operation of graphene field-effect devices, one should control all reliability concerns associated with environmental charge trapping. We study theoretically the effects of external doping on characteristics of graphene field effect devices.

2. Doping and extrinsic doping of graphene

Chemical doping [2, 3] of graphene is an alternative to electrostatic doping, caused by the gate electrode controlled charging. The change in carrier concentration and conductivity of graphene can occur due to external defects, atoms, or chemical molecules. For example, the absorbed gas molecules acting as donors or acceptors. Graphene p-doping occurs even naturally for samples exposed to atmospheric water molecules [4].

Most frequently unintentional doping p-doping is observed, which is believed to originate from adsorbed water molecules due to the water molecules are acceptors. Removing the surface chemical impurities one can significantly reduce unintentional doping. For example, as reported in [5], the intrinsic doping level significantly lowers or vanishes under heating or when placing the graphene under vacuum and pumping for an extended time.

The donor-like and acceptor-like surface impurity atoms and interface defect at graphene-oxide substrate can lead to the shift of the charge neutrality point (CNP) of the gate voltage V_{NP}. This effect quantitatively depends on interface trap density of energy states both for acceptor D_{it}^{acc} and donor D_{it}^{don} traps.

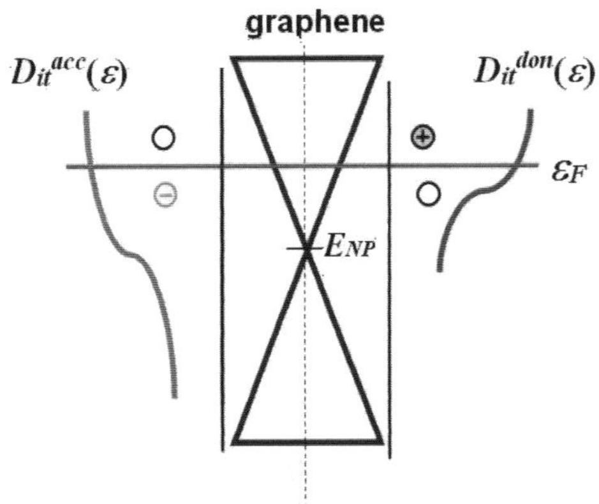

Figure 1. Energy band diagram illustrating chemical doping mechanism. The donor (acceptor) states above (below) the Fermi energy are positive (negative) in thermodynamical equilibrium.

Qualitative mechanism of chemical doping is shown in Figure 1. For example, the donor traps (typically, metal atoms) transfer own electrons to the graphene sheet, remaining in a positively charged state. Generally, the trap charge state and the direction of charge transfer depend on the graphene Fermi energy position with respect to the neutrality point. In its term, the Fermi energy is determined by density of charge traps. A self-consistent problem of computation of Fermi energy, neutrality point gate voltage V_{NP} and charge concentration in graphene will be considered in the next section.

3. Graphene charge densities and Fermi energy as function of gate voltage

A self-consistent calculation of chemical doping effects requires an exact quantitative characterization of charge density and the Fermi energy in graphene. We will rely on this section on the analytical results reported in [6, 7, 8].

3.1. Total and net charge densities in graphene

The amount of charge in the two-dimensional graphene can be characterized by the two types of densities. First, it is the total carrier concentration N_S, which is a sum of the electron (n_e) and the hole (n_h) concentration

$$N_S = n_e + n_h \tag{1}$$

Second, it is the charge imbalance density (net charge density),

$$n_S = n_e - n_h \, . \tag{2}$$

The former is responsible for conductivity of graphene and its I-V characteristics. The latter controls electrostatics and C-V characteristics. Notice, the total charge density practically equals to the net charge density ($N_S \cong n_S$) excepting the vicinity of the charge neutrality point $\varepsilon_F > |k_B T|$.

3.2. Fermi energy as function of gate voltage

The Fermi energy ε_F as function of gate voltage V_G depends on the charge neutrality point gate voltage V_{NP}, and it can be written down as follows [6]

$$\varepsilon_F = \left(m^2\varepsilon_a^2 + 2\varepsilon_a e\left|V_G - V_{NP}\right|\right)^{1/2} - m\varepsilon_a = \frac{V_a}{m}\left[\left(1+2Y\right)^{1/2} - 1\right], \tag{3}$$

where

$$Y = \frac{\left|V_G - V_{NP}\right|}{V_a}, \tag{4}$$

and V_a is defined as follows

$$V_a \equiv m^2\,\varepsilon_a / e, \tag{5}$$

ε_a is a characteristic graphene-oxide constant [6]

$$\varepsilon_a = \frac{\pi\hbar^2 v_0^2 C_{ox}}{2e^2}, \tag{6}$$

$C_{ox} = \varepsilon_{ox}\varepsilon_0 / d_{ox}$ is the gate oxide specific capacitance. The energy ε_a varies in the range from ~1 meV at d_{ox}~ 200 nm and ε_a =4 (SiO_2) to ε_a ~ 0.5 eV at d_{ox}~2 nm and ε_{ox} = 16 (HfO_2). A dimensionless "ideality" factor m in (3) and (5) is defined as

$$m = 1 + \frac{C_{it}}{C_{ox}} \tag{7}$$

where C_{it} is an average interface trap capacitance, which is closely connected with energy density of interface traps $C_{it} = e^2 D_{it}$.

Total carrier concentration can be expressed as an exact relation [8]

$$N_S = n_e + n_h = \frac{\varepsilon_F^2}{\pi\hbar^2 v_0^2} + n_{res}, \tag{8}$$

n_{res} is the residual carrier concentration at the CNP which assumed to be exactly equal to the intrinsic concentration n_i in ideally homogeneous graphene or the total sum of carrier concentration in electron-hole puddles at the CNP. Notice, that the identities $N_S(-\varepsilon_F) = N_S(\varepsilon_F)$ and $n_S(-\varepsilon_F) = -n_S(\varepsilon_F)$ are valid immediately directly under definitions of this variables.

3.3. Total charge as function of gate voltage

Using (3) and (8) $\varepsilon_F(V_G)$ one obtains

$$e\delta N_S \equiv eN_S - en_{res} = C_{ox}\left(\left|V_G - V_{NP}\right| + V_a - V_a\left[1 + \frac{2\left|V_G - V_{NP}\right|}{V_a}\right]^{1/2}\right) = C_{ox}V_a\left[Y + 1 - \left(1+2Y\right)^{1/2}\right] \tag{9}$$

The vicinity around the CNP is characterized by an inequality $Y \ll 1$ ($\left|V_G - V_{NP}\right| < V_a$). The width of this region (~ V_a) is significantly dependent on the thickness of gate insulator and density of interface traps (see (5) and (7)). Given $Y \ll 1$, we have

$$\varepsilon_F \cong \frac{eV_a}{m} Y = e\frac{\left|V_G - V_{NP}\right|}{m}, \tag{10}$$

$$e\delta N_S \cong C_{ox}V_a\frac{Y^2}{2} = C_{ox}\frac{\left(V_G - V_{NP}\right)^2}{2V_a} = \frac{e^2\left(V_G - V_{NP}\right)^2}{\pi m^2\hbar^2 v_0^2}, \tag{11}$$

Charge concentration in this region is dependent on the Plank constants, revealing (11) as a quantum result. This is explained by the quasi-classical approximation is failed near the charge neutrality point, where the carrier's de Broglie wavelength is not small. In contrast, given $Y \gg 1$ ($|V_G - V_{NP}| >> V_a$), we have almost classical "electrostatic" result similar to that in silicon MOSFETs

$$e\delta N_S = C_{ox} V_a \left[Y + 1 - (1 + 2Y)^{1/2} \right] \sim C_{ox} |V_G - V_{NP}|. \tag{12}$$

Despite the fact that the equation $e\delta N_S = C_{ox}|V_G - V_{NP}|$ is widely utilized as an accurate result, in reality, we always have an inequality $e\delta N_S < C_{ox}|V_G - V_{NP}|$.

4. Charge neutrality gate voltage and chemical doping effects

4.1. Electrostatics equation

The basic electrostatic equation in graphene field effect devices with non-uniform energy spectra of interface traps can be written down as follows [6]

$$V_G = \varphi_{GG} + \varepsilon_F + \frac{e^2 n_S}{C_{ox}} + \frac{e^2}{C_{ox}} \left[\int_{-\infty}^{\varepsilon_F} D_{it}^{(acc)}(\varepsilon) d\varepsilon - \int_{\varepsilon_F}^{+\infty} D_{it}^{(don)}(\varepsilon) d\varepsilon \right] - \frac{Q_f}{C_{ox}}, \tag{13}$$

where, energy ε is reckoned from the graphene CNP, $D_{it}^{(acc)}$ and $D_{it}^{(don)}$ are the acceptor and donor-like trap energy densities correspondingly, which assumed to be generally independent, Q_f is density of fixed (i.e., graphene Fermi independent) charge trapped near the graphene – dielectric substrate interface, $\varphi_{GG} = W_G - W_{graphene}$ is the work function difference between graphene $W_G \cong 4.23\,\text{eV}$ and the gate material. It is explicitly assumed in (13) the occupied acceptors are negative and the unoccupied donors are positive. Equation (13) is valid both for electron and hole charging, i.e., for any sign of Fermi energy, including $\varepsilon_F = 0$. The charge neutrality point position is determined by the gate voltage corresponding to zero Fermi energy $V_{NP} = V_G(\varepsilon_F = 0)$

$$V_{NP} \equiv V_G(\varepsilon_F = 0) =$$

$$= \varphi_{GG} + \frac{e^2}{C_{ox}} \left[\int_{-\infty}^{0} D_{it}^{(acc)}(\varepsilon) d\varepsilon - \int_{0}^{+\infty} D_{it}^{(don)}(\varepsilon) d\varepsilon \right] = \varphi_{GG} + \frac{Q_{acc}^{(-)}(\varepsilon_F = 0) + Q_{don}^{(+)}(\varepsilon_F = 0)}{C_{ox}} - \frac{Q_f}{C_{ox}}, \tag{14}$$

where $Q_{acc}^{(-)}(\varepsilon_F = 0) < 0$ and $Q_{don}^{(+)}(\varepsilon_F = 0) > 0$ are the charges of acceptor and donor traps at the neutrality point. Taking into account (14), the basic equation (13) can be rewritten as follows

$$e(V_G - V_{NP}) = \varepsilon_F + \frac{e^2 n_S}{C_{ox}} + \frac{e^2}{C_{ox}} \int_0^{\varepsilon_F} D_{it}(\varepsilon) d\varepsilon, \tag{15}$$

where the total interface trap is defined

$$D_{it}(\varepsilon) = D_{it}^{(acc)}(\varepsilon) + D_{it}^{(don)}(\varepsilon). \tag{16}$$

Notice it is rather hard in practice to distinguish the acceptor and the donor interface traps. Despite their opposite signs, these two types of traps electrically recharged in the same manner. Interface trap capacitance per unit area C_{it} and interface trap level density $D_{it}(\varepsilon)$ are defined in the following way

$$C_{it}(\varepsilon_F) \equiv e \frac{d}{d\varepsilon_F} \left(Q_{acc}^{(-)}(\varepsilon_F) - Q_{don}^{(+)}(\varepsilon_F) \right) = e^2 D_{it}(\varepsilon_F) \tag{17}$$

where the total specific interface trap capacitance is a sum of acceptor and donor state capacitances.

4.2. Graphene charge as function of chemical doping

Let us calculate the density of graphene charge at $V_G = 0$ in a simple case of Fermi-energy independent oxide-trapped charge Q_f. This means that we set the interface trap charge (not density of states!) equals to zero at $V_G = 0$ or it is included in the fixed charge Q_f. In this case, we obtain the explicit equation for charge concentration in chemically doped graphene at the zero gate voltage

$$e\left|n_S\right| = C_{ox}\left(\left|V_{NP}\right| + V_a - V_a\left[1 + \frac{2\left|V_{NP}\right|}{V_a}\right]^{1/2}\right) = C_{ox}V_a\left[\left|\frac{C_{ox}V_{NP0} - Q_f}{C_{ox}V_a}\right| + 1 - \left(1 + 2\left|\frac{C_{ox}V_{NP0} - Q_f}{C_{ox}V_a}\right|\right)^{1/2}\right], \quad (18)$$

where V_{NP0} is the CNP voltage position without any chemical doping. Figure 2 shows calculated results.

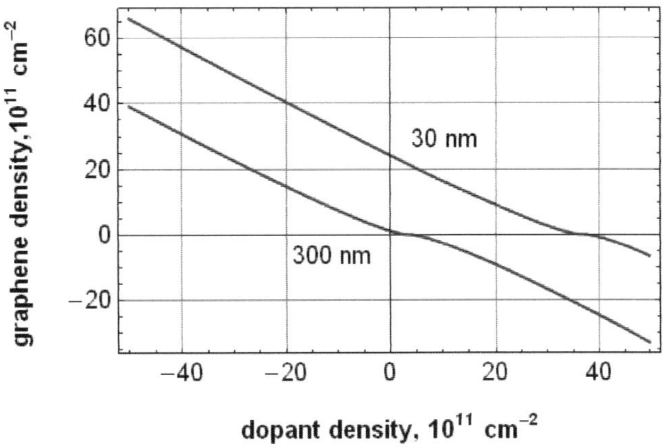

Figure 2. Graphene charge densities as functions of external dopant two-dimensional density for two gate oxide thicknesses 30 nm and 300 nm. Simulations were performed for V_{NP0} = +5V, ε_{ox} = 4 and $D_{it} = 6\cdot10^{12}$ eV^{-1} cm^{-2}.

As can be seen in Fig. 2, the results of chemical doping depend strongly on capacitance (i.e., on the gate oxide thickness and dielectric constant) of gate insulators. Moreover, the efficiency of chemical doping is very sensitive to an initial value of V_{NP0} and energy density of interface traps.

5. Conclusions

We have derived the explicit relations for calculation of carrier concentrations of graphene as functions external charge for different parameters of graphene field effect devices (gate insulator thickness and dielectric constants, interface trap energy density, initial values of neutrality point gate voltages).

References

[1] Novoselov K, Geim A and Morozov S 2005 *Nature* **438** 197
[2] Avouris P 2010 *Nano Lett.* **10** 4285
[3] Farmer D B et al 2009 *Nano Lett.* **9** 388
[4] Schedin F et al 2007 *Nature Materials* **6** 652
[5] Lafkioty M et al 2010 *Nano Lett.* **10** 1149
[6] Zebrev G I 2011 *Phys. and Applications of Graphene – Theory* 475
[7] Zebrev G I et al 2013 *IEEE Trans Electron Dev.* **60** 1799
[8] Zebrev G I et al 2016 *Graphene Science Handbook: Six-volume set (CRC Press)*

IOP Conf. Series: Materials Science and Engineering **475** (2019) 012014 doi:10.1088/1757-899X/475/1/012014

Thermal resistance of 980 nm high-power laser diodes assembled at C-mounts and F-mounts

V V Bezotosnyi[2], O N Krokhin[1,2], V A Oleshchenko[2], V F Pevtsov[2], Yu M Popov[1,2] and E A Cheshev[1,2]

[1]National Research Nuclear University MEPhI, Kashirskoe shosse 31, 115409, Moscow, Russia
[2] Lebedev Physical Institute of the Russian Academy of Sciences, Leninsky Prospect 53, 119991, Moscow, Russia

Corresponding author's e-mail address: victorbe@sci.lebedev.ru

Abstract. We present the experimental results regarding the thermal resistance measurements for 100 μm stripe laser diodes emitting in 980 nm spectral range. The dependence of laser thermal resistance on pumping current was measured for series of identical laser chips with resonator length 4 mm and stripe width 95 mkm assembled directly at copper C-mounts and F-mounts without thermo-compensating submounts inside the output power range 10 W and 15 W accordingly. The appropriate method of spectral envelope peak definition was used for output power range 5-15 W in which we observed spectral peculiarities associated with appearance and competition of several peaks at the spectral envelope. We have proposed an explanation of the obtained dependency based on their correlation with the dependence of the total efficiency of the laser on the pump current.

1. Introduction

Higher total efficiency, higher output power and brightness of high-power laser diodes are the aims on the way to expand the fields of their applications first of all in materials processing as well as navigation and communications [1-3]. Reliable output power around 15 W and even 20 W from 100 μm wide stripe now not seems a dream. We observed quite stable operation of several 980 nm 100 μm stripe 4 mm resonator length chips assembled in our laboratory directly at copper F-mounts during several hours at 20 W CW and got for such selected samples reliable CW operation at 15 W output power without decrease during 100 hours tests [3]. At the base of thermal transfer calculations [4,5] and practical experience of assembling several hundreds of such high-power laser chips at different type heatsinks and submounts we realized clearly that one of the key tasks is to insure efficient heat transfer via the first interface, between the laser chip and the heatsink. So, in order to make the next step towards the increase of output power we need to increase the efficiency of heat transfer at the first step from laser active region to a heatsink because thermal energy is concentrated in quite thin quantum well and quite narrow stripe. At the way to the heatsink the heat flow must propagate in solder layer having a thickness around 3-4 μm and through two interfaces of solder with laser chip and with the heatsink, The uniformity and high quality of bulk solder layer and these 2 interfaces are the key technological tasks. After passing the solder layer, thermal flow density becomes much lower. At 20 W output optical power extremely high heat density around 6-7 GW/ cm^3 is generated inside the active region under heat flux more than 6 kW/ $см^2$ .This CW heat flow must be efficiently removed into a heatsink. At the moment several ways for forced cooling in such extremely hard conditions are implemented – thermoelectrical cooling, microchannel liquid cooling and heat tube or heat camera cooling. The different particular types of forced cooling can be optimal for different cooling schemes, power ranges and heatsinks geometry.

2. Thermal resistance of the laser diode

One of the basic parameters widely used for evaluation the cooling efficiency is thermal resistance. For not-emitting electronic components, it is the ratio of its temperature rise to power supply input rise. For light-emitting devices, especially such as laser diode-with high efficiency, more than 50 % of electrical supply power is converted to optical emission and acts as radiation cooling mechanism of the laser chip. Thermal resistance R_{therm} of laser diode is defined by expression (1).

$$R_{therm} = \Delta T / \Delta P_{therm} \quad (1)$$

where ΔT -the laser chip active layer temperature rise, ΔP – the increase of electrical power supplied by a laser driver, $P_{therm} = IU - P_{opt}$, I – pumping current, U- voltage, P_{opt} – emitted optical power.

In our recent work [6] thermal resistance dependence on pumping current was measured for high-power range up to 15 W CW for 980 nm laser chips assembled at F –mount heatsinks.

3. Experimental results

We repeat here the main result at Fig.1 for reader convenience. First of all, R_{therm} is obvious not-linear and quite complex function of pumping current due to non-linear dependence of peak wavelength against pumping current. Curve A at Fig.1 is added for comparison with the basic case of linear spectral peak wavelength dependence. Unfortunately, in [6] we had possibility to measure only small lot of 3 samples assembled on F-mount heatsinks.

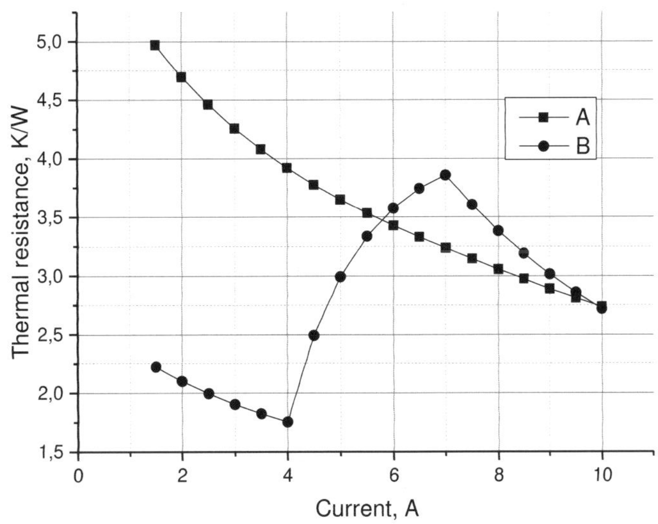

Figure 1. The thermal resistance of 980 nm laser assembled at F-mount heatsinks.
A–"ideal curve" for linear peak wavelength dependence, B- experimental results.

So, to ensure the reliability of the results obtained for F-mounts, here we present similar results for a uniform lot of 10 samples of identical type chips from one gel-pack assembled at C-mounts. The corresponding results on peak wavelength dependence for typical 3 samples from that lot of 10 lasers assembled at C-mounts are presented at Fig. 2. We can say that at pumping currents 1-4 A the peak

IOP Conf. Series: Materials Science and Engineering **475** (2019) 012014 doi:10.1088/1757-899X/475/1/012014

wavelength increase in a linear manner with a slope less comparable with current range 5 -10 A. These results are in principle similar regarding the laser samples assembled at F-mounts.

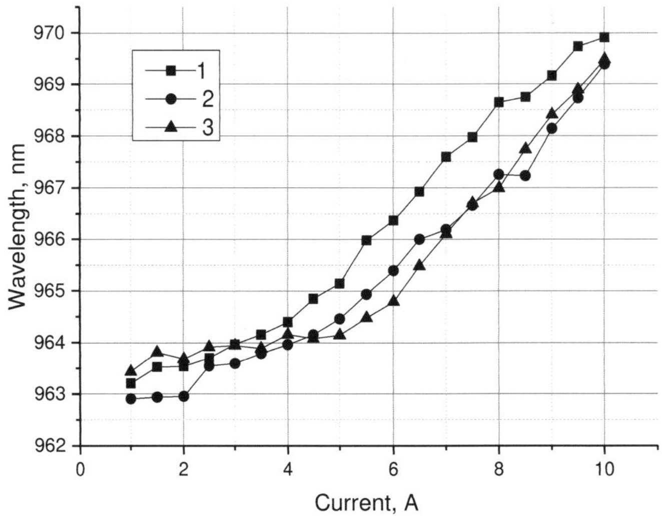

Figure 2. The peak wavelength of 980 nm 3 laser samples assembled at C-mount heatsinks.

The main difference between C-mount and F-mount samples is obvious. For F-mounts in the current range up to 10 A, that we observed 3 main different slopes of peak wavelength dependence on pumping current and for C-mounts in the same current range up to 10A only 2 pronounced slopes. But in both cases the nominal current at which we observed the change of initial slope of peak wavelength dependence was the same, around 4-5 A. The pumping current I_{max} at which we observe the maximum of total efficiency η_{tot} according to our simple analysis was the same - (2), (3) and (4).

$$J = \sqrt{I_{th} U_{th} / R_s} \tag{2}$$

$$I_{\max} = I_{th} + J \tag{3}$$

$$\eta_{tot}^{\max} = \frac{\eta_d J E_g}{I_{\max}(U_{th} + R_s J)} \tag{4}$$

In our opinion, the similar results obtained for 3 tested lasers at F-mounts and 10 lasers at C-mounts present the evidence that not monotonous dependence of thermal resistance on pumping current is a regular and quite important feature of laser diodes. We must keep it in mind in the analysis of several important laser diode characteristics associated with the influence of heat generated inside the active layer. At high currents exceeding 6-10 I_{th} thermal resistance continue to decrease with higher pumping due to the decrease of series resistance R_s. This feature is the same for total efficiency, at very high pumping currents it continues to decrease with lower slope also due to a decrease of R_s.

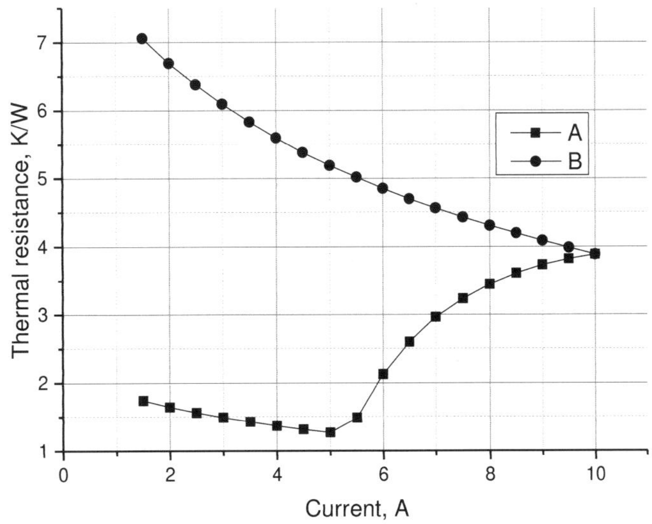

Figure 3. The typical thermal resistance of lasers assembled at C-mount heatsinks. A–"ideal curve" for linear peak wavelength dependence, B- experimental results.

The main features of thermal resistance measured for C- mounts is clear in Fig,3. The curve exhibit the decrease in current range from I_{th} to I_{max}, increase at higher currents and finally coincide with "ideal" curve for linear peak wavelength dependence at value 3.9 K/ W.

4. Conclusions

We carefully examined spectral characteristics of laser diodes in power range up to 10 W CW for C-mount samples and up to 15 W CW for F-mount samples. Spectral peculiarities were observed at output power more than 5-6 W CW for both types of samples. We also measured non- monotonous dependence of emission wavelength on pumping current for both types of samples. We overcome the problem of correct spectral maximum definition by the simple method proposed in [6]. Our experimental data indicate that for identical laser chips are emitting in the spectral range 980 nm and assembled at F-mount as well as at C-mount heatsinks, thermal resistance is also a non-monotonous and quite complex function of pumping current. If we assume the "ideal" linear dependence of emitting wavelength maximum on pumping current, thermal resistance decreases monotonously according to back-proportional law. Real experimental data for both types of samples indicates that thermal resistance correlates with the dependence of total efficiency on pumping current. In the current range from threshold I_{th} to a current corresponding to the maximum of total efficiency I_{max}, thermal resistance decreases in accordance with the increase of total efficiency because more part of pumping power is emitted from laser chip outside. At higher pumping currents we observed thermal resistance increase due to saturation and subsequent decrease of total efficiency at pumping above the current I_{max}.

It means that we can adequately use the definition "thermal resistance" to evaluate the quality of laser chip assembling and effectiveness of different cooling schemes only if we compare its value at the same pumping current. Formally for pumping currents 4-5 A close to I_{max}, we obtained the values of R_{therm} around 1.75 K/ W for F-mount and 1.25 K/ W for C-mount samples. These values are similar and even better than the best-published results for the same samples assembled at composite diamond –copper sub mounts with high thermal conductivity and efficient microchannel cooling [2]. At the same time, we suppose to compare the values at pumping current 10 A more adequate and realistic, where the appropriate values are 2.75 K/ W for F-mounts and 3.9 K/ W for C-mounts. These values are slightly more than 1.5-2 K/ W achieved in [2] but better than 4.25 K/ W obtained in [7]. We believe that managed to get such not bad results using the technology of direct bonding of laser chips to copper heatsinks due first of all to the improvement of interface quality and appropriate improvement of the heat transfer efficiency at interfaces between the laser chip and the heatsink.

Acknowledgements
This work was performed with the support of Program to improve the competitiveness of MEPhI (Agreement No. 02. a03.21.0005).

References
[1] Cruimp P, Decker J, Winterfeldt M et al 2015 *Proc. of SPIE* **9348** 93480D-1
[2] Crump P, Wenzel H, Erbert G, Trankle G 2012 *Proc. Of SPIE* **8241** 824120U
[3] Bezotosnyi V V, Krokhin O N, Oleshchenko V A et al 2015 *Quantum Electronics* **45(12)** 1088
[4] Bezotosnyi V V, Krokhin O N, Oleshchenko V A et al 2014 *Quantum Electronics* **44(2)** 145
[5] Bezotosnyi V V, Krokhin O N, Oleshchenko V A et al 2014 *Quantum Electronics* **44(10)** 899
[6] Bezotosnyi V V, Krokhin O N, Oleshchenko V A et al 2016 *Quantum Electronics* **46(8)** 679
[7] Li X, Zhang Y et al 2012 *IEEE Trans. Compon. Packag. Tehnol.* **2(10)** 1592

IOP Conf. Series: Materials Science and Engineering **475** (2019) 012015 doi:10.1088/1757-899X/475/1/012015

Investigation of semiconducting materials for magnetic field sensors in strong magnetic fields under cryogenic temperatures

I Vasil'evskii[1], A Vinichenko[1], M Strikhanov[1], N Kargin[1], T Kuech[2], K Rogacki[3], I Bolshakova[4], Ya Kost[4], A Moroz[4], F Shurygin[4], A Vasyliev[4] and M Radishevskiy[4]

[1]National Research Nuclear University MEPhI, 31 Kashirskoe Shosse, Moscow, 115409, Russian Federation
[2]University of Wisconsin-Madison, 1848 University Avenue, Madison, 53726, Wisconsin, USA
[3]International Laboratory of High Magnetic Fields and Low Temperatures, 95 Gajowicka Street, Wroclaw, 53-421, Poland
[4]Lviv Polytechnic National University, 12 Bandera Street, Lviv, 79013, Ukraine

Corresponding author's e-mail address: inessa@mail.lviv.ua

Abstract. Influence of strong magnetic fields $B > 3$ T under cryogenic temperatures (1.5 ÷ 4.2) K on the signals of Hall sensors based on single-crystal whiskers (InSb, InAs) and nano-sized heterostructures (InSb/i-GaAs, InAs/i-GaAs) have been investigated. There are distinct Shubnikov – de Haas oscillations for whisker-based sensors, whereas heterostructures-based sensors demonstrate the linear field dependence of the output signal. This difference explained by the higher concentration of structure defects in the heterostructures. The derived results confirm the operability of Hall sensors based on indicated heterostructures for the magnetic fields diagnostics in the temperature and field conditions of modern particle accelerators and fusion reactors.

1. Introduction

Modern charged particle accelerators and plasma confinement systems of experimental fusion reactors like JET, ITER, DEMO tokamaks and W7-X stellarator, etc. as well, call for highly accurate magnetic field measurements.

Reaching this goal for such applications is complicated by the extreme operating conditions including strong magnetic fields (1 ÷ 8) T, wide temperature range (4.2 ÷ 600) K, and irradiation by fluxes of high energy neutrons, electrons and gamma quanta.

The authors have already developed the radiation resistant semiconductor materials for the magnetic field sensors, whose workability were demonstrated in the hard radiation environment of the neutron irradiation up to fluences of $F = 10^{18}$ n·cm^{-2} [1]. The studies have revealed that such indium containing III-V semiconducting compounds as indium antimonide (InSb) and indium arsenide (InAs) show great promise for application in magnetic field sensors intended for operation at high neutron fluxes. The methods for enhancing their radiation resistance including the metallurgical doping during the material growth, as well as the directed radiation modification, were studied. These methods allow one to obtain the semiconductor materials featuring the minimal drift of the sensors characteristics during neutron irradiation, and this drift can be easily corrected by the control electronics.

However, the use of such materials for the high-precision magnetic field measurements in fusion reactors and particle accelerators establishes new requirements for the sensor materials, particularly, the linearity of output signals at the strong magnetic fields and the cryogenic temperatures as well, while maintaining the sufficient magnetic sensitivity.

It is known that the semiconductor materials with the degenerate electron gas and high mobility of charge carriers demonstrate the oscillations of various thermodynamic and kinetic coefficients (conductivity, heat capacity, Hall constant etc.) at the strong magnetic fields and low temperatures. The output signal nonlinearity due to the quantum Shubnikov – de Haas oscillations, limit the accuracy of the magnetic field measurement by the Hall effect sensors based on such materials.

This paper describes the results of the studies on quantum oscillations in indium antimonide and indium arsenide. The samples for investigation were prepared in two different forms with different degrees of structural perfection: the micro-sized monocrystalline InSb and InAs whiskers, and nano-sized thin film InSb/i-GaAs and InAs/i-GaAs heterostructures. Also, the samples featured the different degrees of doping, and the corresponding charge carriers concentrations were in the range of $n = (10^{16} \div 10^{18})$ cm^{-3}.

2. Sample fabrication and experiment procedure

The Shubnikov – de Haas oscillations in the field dependences of Hall effect coefficient in InSb are well known [2, 3]. These studies were carried out on the bulk single crystals.

The goal of this study was to investigate such oscillations in the InAs and InSb thin films that were grown in the heterostructure composition on the semidielectric GaAs substrate to determine the materials parameters at which the quantum oscillations reduce to zero. Moreover, thus, to provide the high accuracy of magnetic field measurements with the help of sensors based on those materials.

In this work the quantum oscillations in single-crystal whickers of InSb and InAs were studied as well to detect the influence of the structure imperfection degree on the amplitude of those oscillations.

The microcrystalline InSb and InAs whiskers were obtained by the deposition from the gas phase by the chemical transport reactions method. The whiskers were grown by free crystallization from the gas phase without any substrate influence, and, therefore, they are characterized by the perfection of the crystalline structure [3, 4]. The mechanical strength of whiskers is close to the theoretical values arising from the interatomic bonds. The formation of crystals with low defects content is also facilitated because they grow at the temperatures well below the melting point. This factor also determines the possibility of the self-cleaning process during the crystal growth. The sizes of whiskers are following: the thickness is (10 ÷ 20) μm, the width is (40 ÷ 60) μm, and the length ranges from one to a few millimeters.

The InSb and InAs whiskers under investigation were plate shaped, and grown along the <100> axis direction, with the side faces indices of {100} and {110}. The electrophysical parameters of InSb whiskers are shown in table 1, while those for InAs whiskers – in table 2. The concentration of charge carriers was targeted by the level of tin doping during whiskers growth.

Table 1. Parameters of InSb whiskers and InSb/i-GaAs heterostructures

Sample No.	Structure	Concentration, n (cm^{-3})	Mobility, μ (cm^2/V·s)	Resistivity, ρ (Ohm·cm)
22	whisker	$2.8 \cdot 10^{17}$	18000	$22 \cdot 10^{-4}$
14	heterostructure	$1.4 \cdot 10^{17}$	16000	$26 \cdot 10^{-4}$

Thin film heterostructures of InSb/i-GaAs and InAs/i-GaAs were obtained by molecular beam epitaxy (MBE) at the NRNU MEPHI Nanocenter and MOCVD method at the Material Research

Science and Engineering Centre of Wisconsin University. The electrophysical parameters of the InSb/i-GaAs heterostructures under study are shown in table 1, those for the InAs/i-GaAs heterostructure – in table 2.

Table 2. Parameters of InAs whiskers and InAs/i-GaAs heterostructures.

Sample No.	Structure	Concentration, n (cm^{-3})	Mobility, μ (cm^2/V·s)	Resistivity, ρ (Ohm·cm)
19	whisker	$5.0\cdot10^{16}$	18000	$68\cdot10^{-4}$
18	whisker	$2.3\cdot10^{17}$	17000	$15\cdot10^{-4}$
17	whisker	$1.7\cdot10^{18}$	900	$4\cdot10^{-4}$
15	MBE heterostructure	$1.4\cdot10^{18}$	8846	$4.8\cdot10^{-4}$
27	MOCVD heterostructure	$3.3\cdot10^{18}$	2970	$15.6\cdot10^{-4}$

The thickness of the InAs active layer in MOCVD method was 100 nm, and for the InSb active layer in the MBE method was about 1 μm. The thin semiconductor layers in heterostructures feature much higher defects content against almost defect free single crystal whiskers. These defects are caused by the misfit dislocations due to the mismatch of lattice parameters of the thin film and the substrate. The metamorphic buffer layers are created to minimize the defects concentration in heterostructures. In the MBE method a 60 nm $In_xAl_{1-x}As$ metamorphic buffer layer was crown, while in the MOCVD method a low-temperature 30 nm InAs layer of was grown.

The studies on single-crystal whiskers and heterostructures were carried out at the International Laboratory of High Magnetic Fields and Low Temperatures (ILHMF&LT) in Wroclaw (Poland), using the MN-1 superconducting solenoid (Oxford Instrument), which provides the range of magnetic field induction of $B = (-14 \div 14)$ T. The field dependences of Hall voltage V_H were recorded at the following fixed temperatures: 1.5 K, 4.2 K, 77 K and 280 K. The B increment during $V_H(B)$ recording was 0.01 T. A Keithley 220 current source was used to sensors power supply, while a Keithley 2000 digital multimeter was used to measure the Hall voltage.

3. Results and discussion

Shubnikov – de Haas effect has been studied quite well both theoretically and experimentally in bulk materials. It is based on the quantization of the electron energy spectrum under the effect of magnetic field. This quantization leads to the appearance of periodic changes in the density of states with the increase in the magnetic field, and correspondingly, the oscillations are observed in both resistance and the Hall voltage field dependences.

The various scattering processes could affect the nature of these oscillations; one should note first the lattice defects, temperature and degree of material doping by impurities.

The present study considers the influence of these scattering factors on the oscillations in Hall voltage in the indium antimonide and indium arsenide samples with different degree of structural perfectness: in whickers 22, 19, 18, 17 and heterostructures 14, 15, 27 samples (see tables 1 and 2).

Figures 1-4 show the field dependences of Hall voltage for whickers and heterostructures with different doping content.

A comparison of field dependences for two InSb samples with different degree of crystalline structure perfectness, but with similar concentrations of carriers is shown in figure 1: an InSb whisker (sample 22, $n = 2.8\cdot10^{17}$ cm^{-3}) is compared with a heterostructure of InSb/i-GaAs (sample 14, $n = 1.7\cdot10^{17}$ cm^{-3}). At the same temperature of $T = 4.2$ K the quantum oscillations are observed in the field dependence for the whisker, but they do not appear for the heterostructure.

Field dependences for three InAs whiskers (samples 19, 18, 17) are given in figures 2 (a, b) and 3 (a). These samples have equal degrees of the structure perfection but differ in the doping level. A weakly doped sample 19 ($n = 5.0 \cdot 10^{16}$ cm^{-3}) shows quantum oscillations. However, with the doping level increasing up to $n = 2.3 \cdot 10^{17}$ cm^{-3} (sample 18) the oscillations amplitude drops from 22% to 8%.

Figure 3 (b) shows the field dependences for the same three single-crystal whisker samples (19, 18, 17) recorded at a higher temperature of $T = 77$ K. One can see that for all whiskers, including the weakly doped sample 19, the oscillations not observed.

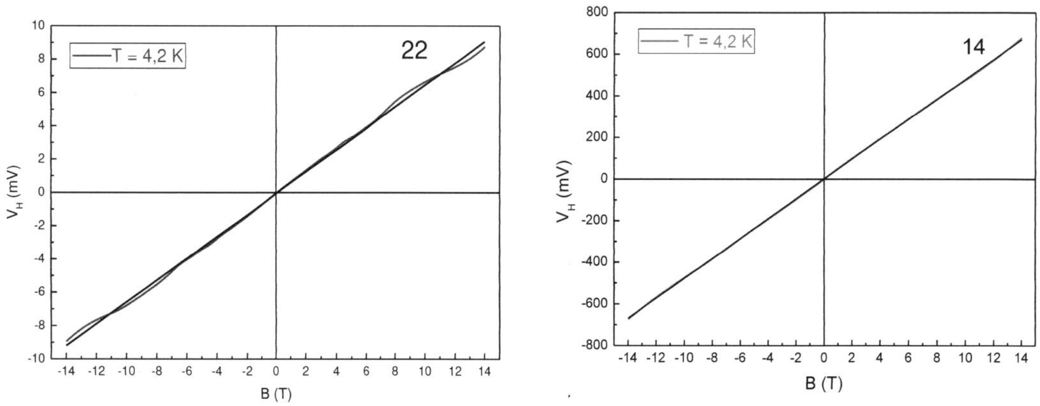

(a) (b)

Figure 1. Field dependences of the Hall voltage:
(a) – InSb whisker (sample 22, $n = 2.8 \cdot 10^{17}$ cm^{-3});
(b) – InSb/i-GaAs heterostructure (sample 14, $n = 1.7 \cdot 10^{17}$ cm^{-3}) at $T = 4.2$ K.

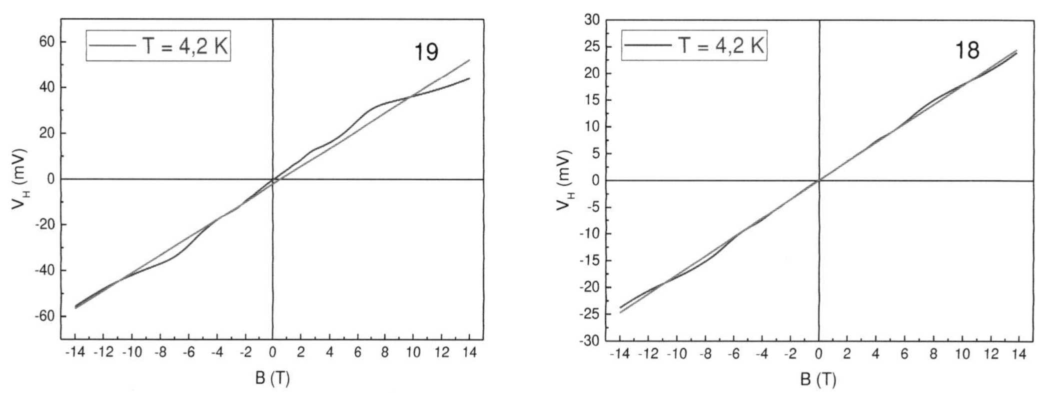

(a) (b)

Figure 2. Field dependences of the Hall voltage for InAs whiskers:
(a) – sample 19, $n = 5.0 \cdot 10^{16}$ cm^{-3}; (b) – sample 18, $n = 2.3 \cdot 10^{17}$ cm^{-3} at $T = 4.2$ K.

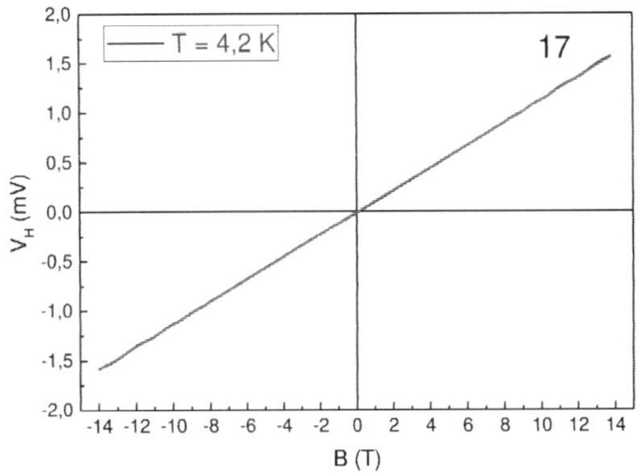

(a)

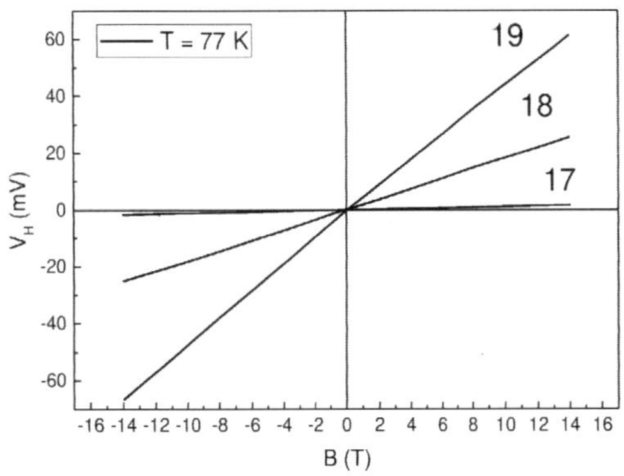

(b)

Figure 3. Field dependences of the Hall voltage:
(a) – InAs whisker (sample 17, $n = 1.7 \cdot 10^{17}$ cm^{-3}) at $T = 4.2$ K;
(b) – whisker samples 19, 18 and 17 at $T = 77$ K.

Figure 4 shows the field dependences of the Hall voltage for two thin films InAs samples being parts of InAs/i-GaAs heterostructures manufactured by different technologies, sample 27 is obtained by MOCVD technique, while sampling 15 – by the MBE. One can see that the both field dependences are quite similar. Only the slope of the lines is slightly different due to small variation in carriers concentration and thickness of the active layer in these heterostructures. However, both are free of the quantum oscillations unlike the weakly doped and more structurally perfect single crystal whiskers.

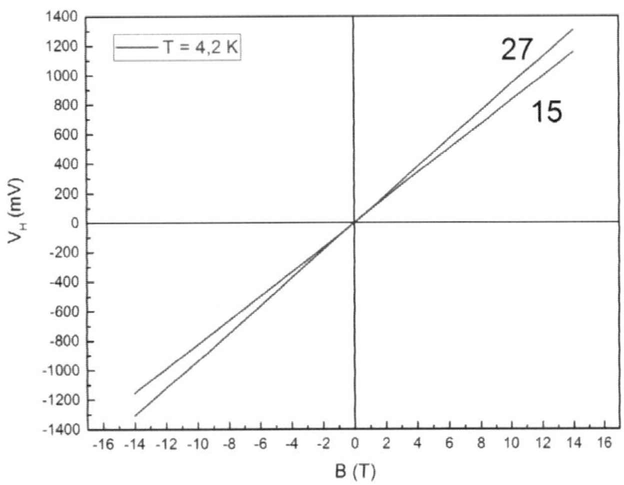

Figure 4. Field dependences of the Hall voltage for InAs/i-GaAs, obtained by two technologies: sample 15, $n = 3.3 \cdot 10^{18}$ cm^{-3}, MBE method and sample 27, $n = 1.4 \cdot 10^{18}$ cm^{-3}, MOCVD method.

The results of these studies reveal the effect of scattering factors like lattice distortion, doping content as well as temperature, on the oscillations in Hall voltage field dependence in samples of heterostructures and single-crystal whiskers of indium-containing semiconductor materials III-V.

These results are of great practical importance. The presence of quantum oscillations in the output signal of Hall effect magnetic field sensors considerably limits the possibilities of increasing the measurement accuracy.

IOP Conf. Series: Materials Science and Engineering **475** (2019) 012015 doi:10.1088/1757-899X/475/1/012015

4. Conclusion

Quantum oscillations of the Hall voltage specific for materials with degenerate electron gas and high carriers mobility have been studied in InSb and InAs semiconductor compounds at strong magnetic fields and cryogenic temperatures. The Shubnikov – de Haas oscillations are observed in structurally perfect single crystal whiskers with a low concentration of carriers. Their amplitude decreases with increase in doping content and carriers concentration, as well as with temperature. These oscillations are absent in thin film heterostructures of InSb/i-GaAs and InAs/i-GaAs over the whole range of concentrations and temperatures at magnetic field ±14 T. This behavior is caused by such scattering factors as structural defects due to the mismatch lattice parameters between the semiconductor material and the substrate. Hence the heterostructures of indium-containing semiconductor materials can be recommended for manufacturing the high-accuracy sensors for measurements of high magnetic fields in the wide range of temperatures, down to cryogen ones.

References

[1] Bolshakova I et al 2015 *Nucl. Fusion* **55** 8 083006

[2] Brandt N and Chudinov S 1982 *Sov. Phys. Usp.* **25** 518–29

[3] Bolshakova I, Moskovets T, Krukovsky S and Zayachuk D 2000 *Mater. Sci. & Engin.* B **69-70** 441–3

[4] Bolshakova I, Koptsev P, Melnyk I, Moskovets M, Krukovsky S and Zayachuk D 2001 *Cryst. Res. and Tech.* **36** 989–96

IOP Conf. Series: Materials Science and Engineering **475** (2019) 012017 doi:10.1088/1757-899X/475/1/012017

Limiting currents of matrix field emission cathodes based on carbon nanotubes

Z Cheng[1], L Sun[1], Z Y Li[2], V Labunov[3], B Shulitski[3], I Kashko[3], D Grapov[3] and N Kargin[4]

[1]Institute of Electron Device & Application, Hangzhou Dianzi University, Hangzhou, China
[2]North-China Electromagnetic Protection Research Institute, Taiyuan, China
[3]Belarusian State university of Informatics and Radioelectronics, Minsk, Belarus
[4]National Research Nuclear University MEPhI (Moscow Engineering Physics Institute), Russia

Corresponding author's e-mail address: labunov@bsuir.by

Abstract. The limiting emission currents of matrix field cathodes based on carbon nanotubes are investigated for their dependence on the cathode working area, as well as the features of the Fowler-Nordheim curves in cyclic measurements of the I-V characteristics under cyclic measurements of CVC. The migration of the thermal overload region of emitters with a temperature of more than 1500 K (with emission in the visible spectral range) on the surface of a matrix cathode is set, as well as two types of divergence of the Fowler-Nordheim curves under cyclic measurements of CVC.

1. Introduction

One of the most noticeable trends in recent years is the advancement of vacuum field emission electronics in the regions of short millimeter waves and the terahertz range [1]. These are, in the first place, devices of microwave electronics – triodes, klystrons, magnetrons, traveling wave tubes for space communication systems, radar, IR observations (visualization and identification), and moreover, for diagnostic purposes - sensing dangerous goods (explosives, drugs), among other uses. The key problem of vacuum microwave electronics is that of the high-current field emission cathodes.

A promising material for creating high-current field emission cathodes are carbon nanotubes (CNTs) [2, 3]. The prospects for the use of CNTs are due to a unique combination of their electrical, chemical, optical, thermal and mechanical properties. Important properties of CNTs are namely their high electrical conductivity and unusually high emissivity due to their quasi-one-dimensional structure with a high aspect ratio. Further advantages of CNTs are a high coefficient of amplification of the electric field by an individual emitter, unattainable in any of the known types of emitters - up to 30,000 rel. units, the record low threshold strength of the emission field for all known types of emitters - at a threshold electron emission current of 1 mA / cm^2 - 0.2 V / μm, record high thermal conductivity of the individual emitter (higher than that of diamond emitters) - 3000 W / (mK). Other advantages of field emission cathodes on the basis of CNTs are resistance to all types of radiation, up to complete destruction of the material, stable operation over a wide temperature range from -130 °C to +300 °C, lack of inertia and low power consumption. However, to date, commercial devices based on CNT field-emission cathodes are not available on the world market.

It is possible to distinguish three characteristic parameters that determine the possibility of using CNT field-emission cathodes for instruments and devices of field emission microwave electronics: a - field emission current density J_{FE}; b - operating voltage U_{FE}; c - stability of emission parameters and durability of CNT field-emission cathodes. In this article, we consider issues related to the problem of obtaining high current density of field emission of CNT field-emission cathodes.

For a number of significant applications in the field of microwave electronics (triodes, klystrons, traveling wave tubes) are needed $J_{FE} \geq 1$ A/cm^2 from the working area of the cathode $S_W \geq 1$ cm^2. The theoretical limit for field-emission cathodes based on nanosized carbon materials reaches $J_{FE} = 10^9$ A/cm^2 [4]. For the matrix of point emitters, even with allowance of space-charge effects, are theoretically achievable $J_{FE} = 10^3$ A/cm^2 [5]. In addition, it was shown in [6] that the screening effect of the space charge in the case of a nanoscale emitter is insignificant up to the limiting emission current densities. However, to date, in practice, even under laboratory conditions, the limit of the emission current density for CNT field-emission cathodes from the cathode working area $S_W \geq 1$ cm^2 is not more than 1 A/cm^2.

In this work, we have studied the limiting emission current as a function of the working area of the matrix CNT field emission cathodes, as well as the characteristics of the Fowler-Nordheim curves with multiple cyclic passes to the region of limiting currents when measuring the current-voltage.

2. Experiment

CNT field emission cathodes were produced by CVD synthesis technology with the catalytic pyrolysis of the hydrocarbon mixtures of ferrocene $(C_5H_5)_2Fe$ with xilene (C_8H_{10}) (0,15 ÷ 0,5 cm^3/min) in the presence of the localized and pre-structured catalyst on the substrate. The content of ferrocene in this "feeding solution" was 0,1% wt %. A flow rate of carrier gas (Ar, N_2) 100 cm^3/min. As a substrate for subsequent CNTs growth, *n*-type silicon wafers with high doping level KES-0, 01 (100) was used. The localized catalyst is a metallic layer of Al\Ni with a thickness of 10 nm and 1 nm, respectively. To prevent interaction of Al with Si the Si-substrate was coated with molybdenum film with a thickness of 200 nm. The topology of the test structure is obtained by lift-off photolithography. Immediately prior to the CNT synthesis, the substrate was subjected to mechanical cleaning and washing in alcohol (ethanol or isopropanol). The topology is a set of simple elements 2×2 μm^2 dots with 100 μm and 5 μm spacing. The active area of test topology was 1,2×3 mm. High reproducibility in combination with low synthesis temperature (650 °C) yielded the cathodes with an emitter diameter 5 – 10 nm, high adhesion to the substrate and a minimum height nonuniformity of nanotubes.

As it is shown in our patent [7], the use of volatile catalyst Fe from $(C_5H_5)_2Fe$ source in combination with the multilayer localized catalyst Mo\Al\Ni, play the role of the activator of metal-organic compound dissociation, which insures the obtaining of the few-wall (2 – 6 walls) CNTs with narrow diameter distribution of low defect density, high adhesion to the substrate and minimum micro- and macro nonuniformity of CNTs height.

Study of the geometric features of the experimental samples of CNT FECs were carried out using a scanning (JEOL 6510) and transmission (JEOL 100CX) electron microscopes. The FE properties of the experimental samples of CNT cathodes were investigated with the field emission scanning microscope (FESM) for spatially resolved measurements (University of Wuppertal, Germany) at a pressure of 10^{-9} with Torr residual gases, and also using a diode-type cell with a fixed spacer (BSUIR, Belarus). The migration processes of the emission region along the cathode surface have been investigated in real time using a field emission scanning microscope (FESM) and a built-in video camera.

3. Results and discussion

3.1 The relationship of the emission current density and the working area of the cathode

Figure 1 (a) shows a micrograph (scanning electron microscopy, SEM) of a single, separately standing element of a CNT field-emission cathodes matrix with a geometric size of 2 μm, distanced from the others by 100 μm. The internal structure of an element consisting of carbon nanotubes 5 – 10 nm in diameter is shown in Figure 1 (b) (transmission electron microscopy, TEM). Figure 1 (c) shows the current-voltage characteristic of a single element of the CNT field-emission cathodes matrix measured in DC mode using a field emission scanning microscope and an anode of 30 μm in diameter. Since the distance between the unit elements is 100 μm, the working area of the emission is determined by the size of a single, separately standing element of the CNT field-emission cathodes matrix. The maximum achievable value of the emission current is obtained for a single element $I_{FE} = 2{,}8 \cdot 10^{-5}$ A, as illustrated in Figure 1 (c)). The dashed line in Figure 1 (a) denotes an area of 5 × 5 μm^2, corresponding to a single element in the matrix structure with a pitch of 5 μm, excluding the screening of the field by adjacent elements (excluding the near-field effect).

The matrix structure of CNT PEC of densely packed single elements with a 5 μm step is shown in Figure 2 (a, b) (SEM). Packing density of unit cells in the cathode matrix is $4 \cdot 10^6/cm^2$. Taking into account the emission current value of $2.8 \cdot 10^{-5}$ measured at the unit cell and the unit packing density of $4 \cdot 10^6$ / cm^2, the estimated expected (potential) density of the field emission current of such a matrix CNT field-emission cathodes $J_{FE} = 2{,}8 \cdot 10^{-5} A \times 4 \cdot 10^6/cm^2 = 110$ A/cm^2.

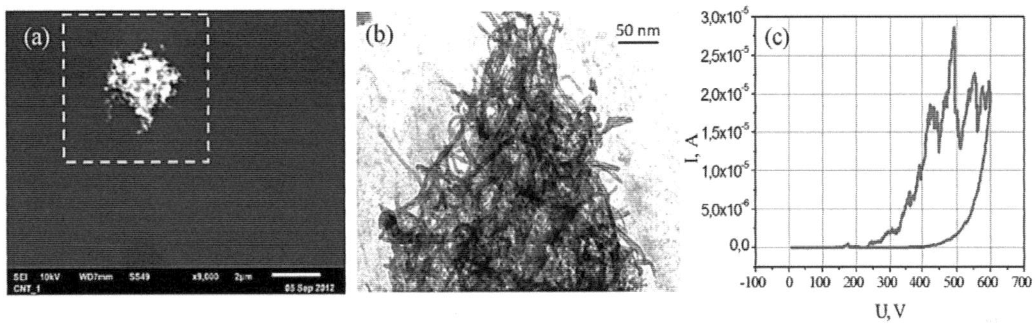

Figure 1. Micrograph (a) scanning, (b) transmission electron microscopy, and (c) current-voltage characteristics of a single element of the matrix CNT cathode distanced from others.

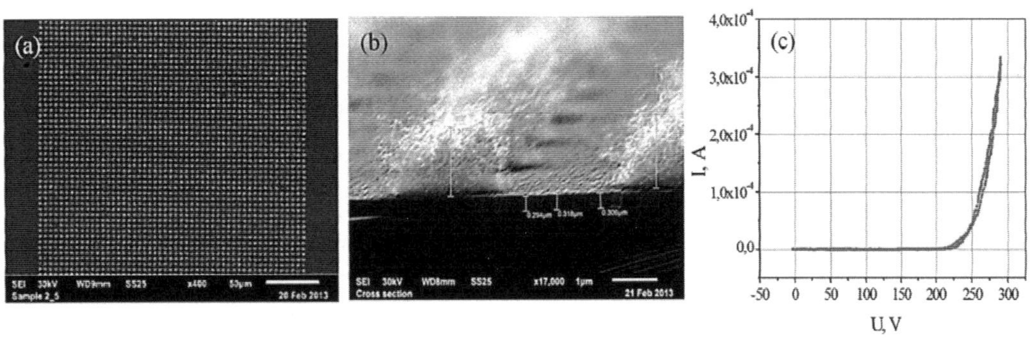

Figure 2. Microphotographs (a, b) of scanning electron microscopy and (c) current-voltage characteristics of matrix CNT cathode.

Figure 2 (c) shows the actual CVC of a matrix cathode with a unit element pitch of 5 μm, measured in DC mode from the working area of the emission $S_W = 1{,}7 \cdot 10^{-4}$ cm^2, corresponding to an anode diameter of 150 μm. The anode-cathode distance is 50 μm, the topology of the sample corresponds to Figure 2 (a, b). As follows from Figure 2 (c), the maximum value of the measured emission current is

I_{FE} = 3,4·10^{-4} A, which corresponds to the estimated current density J_{FE} = 3,4·10^{-4} A × 1,7·10^{-4} cm^2 = 1,9 A/cm^2. As we can see, with an increase in the area of emission of the matrix cathode, the density of the emission current differs sharply from the calculated supposition on the basis of field emission measurements from a small cathode area.

The graph of the J_{FE} with S_W dependence of a matrix cathode with a step of unit elements of 5 μm is shown in Figure 3 (curve 1). As follows from the graph, the dependence of J_{FE} on S_W is linear. The maximum value of J_{FE} = 110 A/cm^2 was received from the working area of the cathode S_W = 2,5·10^{-7} cm^2. With an increase in S_W, the values for J_{FE} fall sharply and with S_W = 1,7·10^{-4} cm^2 we obtain J_{FE} = 1,9 A/cm^2 (Figure 3, curve 1). Consider the possible causes of the fall of J_{FE} with increasing S_W.

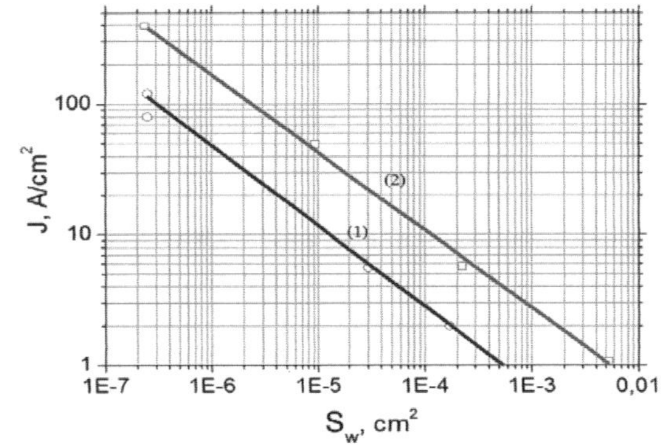

Figure 3. Experimental dependence of the current density on the working area of the CNT FEC

3.2 Analysis of the relationship between the emission current density and the cathode working area

One possible reason for the observed patterns of the dependence of J_{FE} on S_W is the statistical spread of the structural parameters of individual emitters over the area of the matrix cathode, associated with the imperfection of the technology of CVD synthesis of CNT. In the case of matrices with a significant number of tip emitters, to achieve limiting J_{FE}, ideal uniformity (unification) of the structure along the height and diameter of the individual emitters is required, as well as sufficient distance between them and the magnitude of the interelectrode gap. In practice, the ideal unification of the structure of CNTs in the processes of self-organization (growth) is technologically difficult to achieve in all known growth technologies, such as CVD, arc discharge, laser ablation, etc. In all mechanisms of synthesis, it is impossible to avoid heterogeneity in diameter, height of carbon nanotubes, and the distance between them.

The emission current density J_{FE} of field-emission cathodes is described by the modified Fowler-Nordheim expression [8-10]:

$$J_{FE} = \frac{A\beta^2 E_{FE}^2}{\varphi} \exp\left[\frac{-B\varphi^{3/2}}{\beta E_{FE}}\right] \text{ A/cm}^2, \quad (1)$$

where E_{FE} – the intensity of an external electric field (V / cm), defined as $E_{FE} = U_{FE}/d$, where U_{FE} – operating voltage, d – interelectrode gap, φ – the work function of the emitter material (eV), β – a field amplification factor, A = 154 and B = 6830 – constant values. For multi-walled carbon CNTs φ = 5 eV [11, 12].

The most significant variable in the Fowler-Nordheim equation is the value of the field amplification factor β, which determines the degree of transformation of an external macroscopic field into a local nanoscopic field at the end of a nanotube, leading to electron emission. Experimentally β can be determined from the slope of the Fowler-Nordheim graph, which is a graph of the function ln(I/E^2) = f(1/E) and for the case of field emission has the form of a straight line. In this case β is calculated as:

$$\beta = \frac{C\phi^2}{tg\alpha}, \tag{2}$$

where $tg\alpha = d(\ln(I/E^2))/d(1/E)$ – the tangent of the inclination angle of the Fowler-Nordheim curve, φ – the work function of the emitter material and C is the intersection point of the Fowler-Nordheim graph with the ordinate axis.

In the paper [10], the idea of a hybrid field amplification scheme was proposed, according to which the field amplification effect has two components: field amplification factor β (internal factor) and external factor γ, defined by electrostatic screening, connected with the close arrangement of individual CNTs relative to each other (near-field effect). Internal factor (β) is determined by one separately located free-standing CNT without neighbors. It is determined by the geometry of the CNT and can be calculated as $\beta = l/r$ [13], where l – length of CNT, r – curvature radius of the tip of CNT. External factor (γ) can be determined by calculation using numerical methods.

In practice, the decisive influence on the current density of the emission in the array is exerted by an external factor γ [14]. In paper [15] it is shown that field-emission cathodes should provide optimal emission efficiency in the case that the length of CNT (l) is at least 10 times greater than the diameter, and the distance (h) between two neighboring CNTs is at least $2l$. This means that the array must be sufficiently sparse. On the other hand, a significant decrease in the density of the array can lead to a decrease in the current density due to a decrease in the number of emitting CNTs per unit area. Therefore, in order to obtain high efficiency of field-emission cathodes, it is important to choose the optimal ratio of length, diameter and density of CNTs in the array.

In accordance with the Fowler-Nordheim formula [8] considered above, as well as its generalization, the Murphy-Hood formula [16], the values E_{FE} and J_{FE} are related by an exponential relationship. Depending on the region of the curve J_{FE}/E_{FE} increment of E_{FE} at 10 % may lead to an increase of J_{FE} of an individual emitter in 5 – 10 times. At the same time, the value of the statistical variation in height and diameter of individual emitters proportionally changes the required value of the electric field strength E_{FE}, since the height determines the interelectrode gap between the emitter and the anode, and the emitter diameter is the internal field amplification factor β on the individual emitter. Thus, a statistical spread of 10% in height and diameter of individual emitters can lead to a spread of field emission currents from individual emitters by 5 – 10 times. In turn, a change in the distance between the individual tip emitters also contributes to the spread of the value of J_{FE}, since it changes the value of the external field amplification factor γ, associated with the screening of the field with a sufficient approach of the emitters (near-field effect).

The statistical spread of the geometric parameters of neighboring tip emitters (individual carbon nanotubes) can be defined as the microinhomogeneity of the structure. At the same time, with an increase in the working area of the cathode S_W the heterogeneity of the CNTs parameters at different parts of the cathode matrix also begins to play a role, determined by the heterogeneity of the growth conditions of CNT on the extended synthesis zone in the CVD reactor. Such heterogeneity can be defined as macroinhomogeneity. Starting from a certain micro surface area, the microinhomogeneities of the structure are averaged and with an increase in the working area of the cathode matrix, it is macroinhomogeneities that can make the main contribution to limiting the value of J_{FE} measured in practice.

In CVD synthesis processes, one of the key parameters that determine the diameter and growth rate (height) and density of carbon nanotubes are the geometric dimensions and topology of metal catalyst clusters [17]. As a consequence, one of the ways to unify the geometric parameters of CNTs can be use of nanolithography technology to specify the shape, size, and exact localization of metal-catalyst clusters on a substrate. Such work was carried out within the framework of the project CANVAD of the 5th European Framework Program [18, 19]. The topology of clusters was formed using electron-beam lithography. The cathode matrix obtained from the CVD process was an almost perfect structure with a distance of 10 μm (with an accuracy of 1%) of CNT field-emission cathodes with a height of 5 μm (with an accuracy of 4%) and a diameter of 50 μm (with an accuracy of 6%) [18,

19]. At these cathodes, when measured in the DC mode from cathode working areas $S_W = 2{,}2\cdot10^{-7}$ cm^2 and $S_W = 3\cdot10^{-3}$ cm^2 calculated emission current density was obtained, respectively, $J_{FE} = 400$ A/cm^2 and $J_{FE} = 1$ A/cm^2. If we denote on the graph in the coordinates J_{FE}/S_W position of the obtained measurement results, then we observe the same pattern associated with a sharp fall J_{FE} with increasing S_W (Figure 3, curve 2).

Thus, even with the use of nanolithography technology, it is impossible to avoid a characteristic drop-dependence of J_{FE} on S_W. It is interesting to note that in the article [18], a summary graph of the distribution of the measurement results of CNT field-emission cathodes is presented on the basis of an analysis of 14 publications by different authors. The same characteristic feature of the graph is observed: a drop in the values J_{FE} with increasing of values S_W. In particular, when S_W corresponds to 10^{-4} cm^2 they report receiving $J_{FE} = 6$ A/cм^2 [20], however when S_W is 10^{-1} cm^2 the maximum achievable is only $J_{FE} = 10^{-3}$ A/cm^2 [21], which correlates with the results obtained by us (Figure 3, curves 1 and 2).

3.3 Interrelation of the emission current density and heterogeneity of the CNT field-emission cathodes structure

Let us consider possible mechanisms of J_{FE} restriction of CNT field emission due to the statistical spread of geometric parameters. Almost all the advantages of carbon nanotubes for field-emission cathodes applications listed at the beginning of this article (see Introduction) refer to single-wall CNTs of small diameter (nanometer units) with metallic conductivity type. However, as a rule, real individual emitters are multi-walled carbon nanotubes, which are characterized by low electrical conductivity, a significant defectiveness of graphene walls and the problem of mechanical, electrical and thermal contact with the substrate [22]. These problems can lead to strong heating of emitters during the passage of field emission currents (Joule heat), especially on structural defects and in the area of poor contact with the substrate. In addition, as the voltage increases due to the statistical spread of the structural parameters, individual emitters on the working area of the cathode matrix are loaded non-uniformly. The temperature of the more emitters (with a larger height and smaller diameter) is rising. The exceeding by I_{FE} of some critical limit leads to thermal destruction of the overloaded individual emitters and transfer of the load to neighboring emitters. At the same time, from our point of view, several variants of the development of events are possible. If the loaded emitters have shielded neighboring emitters (due to the microinhomogeneities of the cathode structure), then after their destruction, these neighboring emitters are activated, and the effective emission region retains its localization. Otherwise, the emission migrates to other regions of the emitter matrix, determined by the macro-inhomogeneities of the cathode structure.

In the article [23], in which the distribution of emitted electrons by energy was studied, it was established that multi-walled CNTs in the emission process could be heated up to 2000 K without significant loss of emission properties. However, the authors have worked in the field of low density field emission currents. In the case of high J_{FE}, it may be possible that the temperature rises above 2000 K, since the destruction of the emitter carbon material recorded by us at the emission limit currents can correspond to a temperature of 4500 K. High temperatures make it possible to record the localization of the loaded emission regions by the characteristic glow of heated emitters in the visible spectral range, starting from temperatures above 1500 K [23]. We investigated these processes in real time using a field emission scanning microscope (FESM) and a built-in video camera.

Figure 4 (a, b) shows microphotographs of the emission region of CNT field-emission cathodes for different values of the operating voltage obtained with the built-in video camera. When the working voltage is changed, the area of intense glow also changes its shape. The luminescence region can be defined as the area of effective emission, and the area occupied by it is the effective emission area. Obviously, the effective emission area can be very different from the working area of the cathode S_W.

Figure 4 (c) shows the Fowler-Nordheim dependence for a sample of CNT PEC for multiple passes (rise and fall) of the operating voltage. A characteristic feature of these Fowler-Nordheim curves is the presence of saturation of the emission current, known as the effect of emitter current saturation [18].

Saturation occurs at the emission currents $I_{FE} \approx 10^{-6} - 10^{-5}$ A (in our case, this corresponds to current densities of $J_{FE} \approx 5 - 50$ mA/cm^2) and correlates in time with the appearance of the luminescence effect in the emission area. The saturation effect was recorded by us on all types of samples studied when $J_{FE} > 5 - 50$ mA/cm^2 was exceeded and was considered in detail in our work [24]. Another observed feature of the Fowler-Nordheim curves is the splitting of the curves with multiple cyclic passes to the area of the maximum emission currents (Figure 4 (c)).

An example of the behavior of another type of sample, with a high macroinhomogeneity of the structure of the matrix cathode, is shown in Figure 5. Microphotographs, as depicted in Figures 5 (a-d), show effective emission regions, determined by the characteristic luminescence of overloaded emitters. As the voltage rises, migration of the luminescence area over the cathode working area is observed due to the high macroinhomogeneity of the cathode structure. The features of the behavior of a sample of the first type, in Figure 4 (a-c) can be attributed to the case of fairly homogeneous samples of CNT field-emission cathodes with a low macroinhomogeneity of the structure. For a sample of the second type, such as that depicted in Figure 5 (a-e), we can assume a sufficiently high macroinhomogeneity of the cathode structure.

3.4 Features of the Fowler-Nordheim dependencies in cyclic CVC measurements

The difference in the character of the heterogeneity of the structure is reflected in the Fowler-Nordheim curves. Depending on the type of prevailing inhomogeneity, there is a different character of the divergence of the Fowler-Nordheim curves for multiple passes of the CVC measurements. In the case of sufficiently homogeneous samples of CNT field-emission cathodes with a low macroinhomogeneity of the structure, the tangent of the curve slope to the coordinate axes is observed to change, as in Figure 4 (c), in the other case, the tangent of the slope angle remains unchanged, as in Figure 5 (e). Let us try to find an explanation for these features.

As follows from formulas (1) and (2), the tangent of the slope angle tgα in the Fowler-Nordheim coordinates is proportional to the ratio of the work function of the material φ and the field amplification factor β (tg$\alpha \approx \varphi/\beta$). The work function depends on the sorption-desorption processes of the residual gases at the end face of the emitter, as well as the temperature of the emitter [25]. The field amplification factor β is determined mainly by the geometric parameters (the apex ratio) of the emitter. The measurements were carried out using field emission scanning microscope (University of Wuppertal, Germany) at a residual gas pressure of 10^{-9} Torr, which minimizes the role of sorption-desorption processes of residual gases at the end face of the emitter at multiple passes. A more significant contribution can be made by the work function of the material φ, which is essentially temperature dependent. With the onset of the transition to the thermal overload of the emitter after the inflection point (the beginning of the process of saturation of the emission current due to the drop in the large potential on the emitter, the heating by Joule heat, the appearance of intense luminescence in the emission region), the most effective individual emitters (carbon nanotubes with a high aspect ratio) are destroyed. However, this reduces the work function of the output φ, which compensates for the decrease in the amplification factor β, and the tangent of the slope of the curve of tgα remains unchanged. When the operating voltage returns backward, the work function of the output φ restores its value, however, the amplification factor β is significantly reduced due to the destruction of active emitters during the high-temperature heating period, which leads to a change in the tangent of the slope angle tgα below the inflection point and explains the characteristic form of the Fowler-Nordheim curves diverging downwards, shown in Figure 4 (c).

In the other case, in the presence of high macroinhomogeneity, with repeated passes and returns to the region of low stresses (and, correspondingly, temperatures), the slope tangent tg α remains unchanged. Taking into account the stability of the work function φ in the low-temperature region (self-heating of the emitter being negligible), it can be concluded that the field amplification factor β remains practically unchanged. This can correspond to the case when the effective emission region shifts (migrates) over the cathode working area with the inclusion of new cathode regions with the same microstructure of the emitters and the field amplification factor β, respectively, after each passage, yet with an increased interelectrode gap. On micrographs, we see the shift of the luminescence region (the area of effective emission) to new regions of the working area of the cathode matrix, displayed in Figure 5 (a-d). It can be assumed that the migration of effective emission regions allows maintaining the stability of the field amplification factor β for multiple passes in the case of high macroinhomogeneity of the structure of the matrix cathode, in Figure 5 (e). The areas of effective emission change their localization, yet the total size of the effective emission area can remain constant.

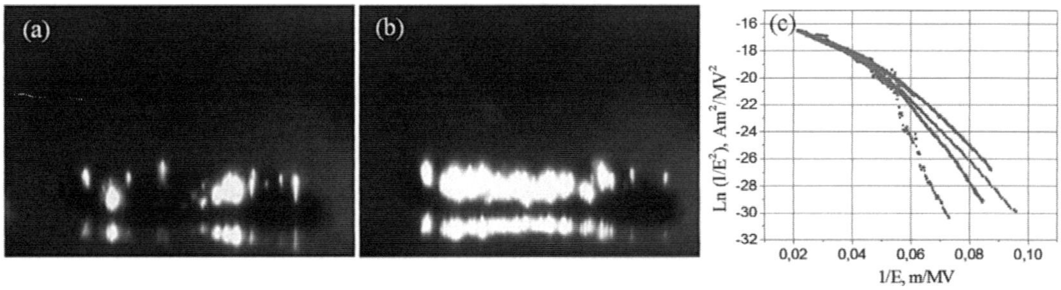

Figure 4. Microphotographs of the working area of the emission at different values of the field emission current (a-b) and the Fowler-Nordheim dependence (c) for a sample of CNT field-emission cathodes with low macroheterogeneity of the cathode structure.

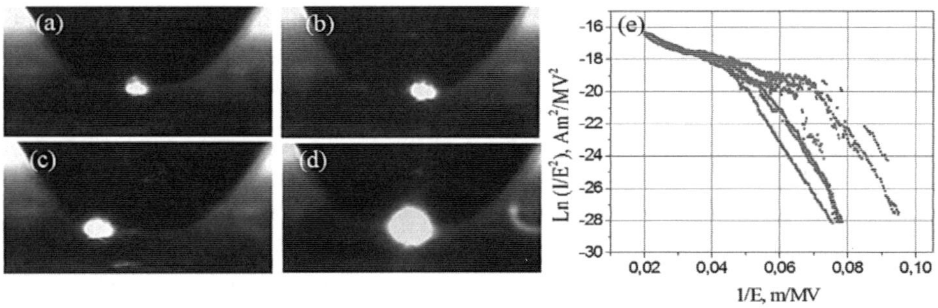

Figure 5. Microphotographs of the working area of the emission at different values of the field emission current (a-d) and the Fowler-Nordheim dependence (e) for a sample of CNT field-emission cathodes with high macroheterogeneity of the cathode structure.

Thus, the characteristic feature of the divergence of the Fowler-Nordheim curves found by us at multiple passes of CVC-measurements can be explained by migration of the effective emission region along the cathode working surface, recorded by the characteristic luminescence of an overloaded current region. As in the cases of migration and fixed emission areas, a common feature is a significant difference between the effective emission area and the cathode working area S_W, which can be considered as one of the main reasons for limiting the practical values of the density of the autoemission current of CNT field-emission cathodes. It is the micro- and macroinhomogeneities, in the first place, that determine the characteristic features of the field electron emission processes and the limits of the emission current densities achievable in practice on the matrix of the pointed CNTs field-emission cathodes. At the same time, intense luminescence in the region of effective emission before current disruption testifies to the thermal character of irreversible degradation and destruction of emitters, as can be seen from the strong stratification of the Fowler-Nordheim curves after the rise to the region of limiting currents.

4. Conclusions

The limiting emission currents of matrix field cathodes based on carbon nanotubes are investigated depending on the cathode working area, as well as the features of the Fowler-Nordheim curves in cyclic measurements of CVC. The migration of the thermal overload region of emitters with a temperature of more than 1500 K (with emission in the visible spectral range) on the surface of a matrix cathode is determined.

It was found that the measurement of current-voltage characteristics in the mode of multiple passes to limit the emission currents has several options of emission development processes. In the first case, when returning to the low-current region, the value of the field amplification factor β drops, which can be associated with the destruction of the most efficient emitters in the field of effective emission as a result of thermal overload. In another case, the field amplification factor β retains its value, which may be due to the migration of the effective emission area on the surface of the matrix cathode.

Thus, it is possible to single out the key problem of obtaining a high limiting current density from matrix cathodes based on carbon nanotubes. The problem is related to the micro- and macroinhomogeneities of the cathode structure, which are the reason for limiting the area of effective emission in comparison with the cathode working area. In order to solve this problem, it is required to search for a synthesis technology that is unified by geometric parameters (height, diameter, distancing) of defect-free CNTs over a large area of the substrate surface (more than 1 cm^2) with low micro- and macroinhomogeneity of the structure, both. To solve this problem would open the way to getting high emission current densities for matrix field emission cathodes based on carbon nanotubes with a density of emission current of up to several thousand amperes per cm^2 and to the creation, on the basis of CNT field-emission cathodes of a new generation of instruments and devices of high-current field emission microwave electronics.

Acknowledgments
We wish to thank professor G. Muller (University of Wuppertal, Germany) for his help in arranging the measurements on the field emission by field emission scanning microscope.

References

[1] Srivastava Anurag 2015 *Europ. J. of Adv. in Eng. and Technol.* **2(8)** 54
[2] Chernozatonskii L A et al 1995 *Chem. Phys. Lett.* **233(1)** 63
[3] Yeletsky A V 2002 *UFN* **(172)4** 401
[4] Fursey G N, Baskin L M , Glazanov D V et al 1998 *J. Vac.Sci. Technol. B* **16(1)** 232
[5] Zhu W 2001 *Vacuum Microelectron.* 416
[6] Pavlov V G 2004 *J. Tech. Phys.* **74(12)** 72
[7] Saurov A N et al 2012 *Patent of the Eurasian Patent Office* EA016557
[8] Fowler R H and Nordheim L 1928 *Proc. Roy. Soc. Lon.* **A119** 173
[9] Burgess R E, Kroemer H and Houston J M 1953 *Phys. Rev.* **90** 515
[10] Kim D, Bouree J E and Kim S Y 2006 *Appl.Phys. A* **83** 111
[11] Gröning O, Küttel O M, Emmenegger C et al 2000 *J.Vac. Sci. Technol. B* **18** 665
[12] Bonard J M, Croci M, Arfaoui I et al 2002 *Diamond Relat. Mater.* **11** 763
[13] Mahner E, Minatti N, Piel H and Pupeter N 1993 *Appl. Surf. Sci.* **67** 23
[14] Wang T, Reece C E and Sundelin R M 2003 *J.Vac. Sci. Technol. B* **21** 1230
[15] Egorov N V and Almazov A A 1999 *Vacuum* **52** 295
[16] Murphy E L and Good R H 1956 *Phys. Rev.* **102** 1464
[17] Bulyarskiy S V 2011 *Mater. Electron. Eng* **2(54)** 38
[18] Milne W I, Teo K B, Minoux E et al 2006 *J. Vac. Sci. Technol. B* **24(1)** 1074
[19] Minoux E et al 2007 *Proc. of the IEEE Int. Conf. Nanotechnology*
[20] Zhu W, Bower C, Zhou O et al 1999 *Appl. Phys. Lett.* **75** 873
[21] Murakami H, Hirakawa M, Tanaka C and Yamakawa H 2000 *Appl. Phys. Lett.* **76** 1776
[22] Kataura H, Kumazawa Y, Maniwa Y et al 1999 *Synth. Met.* **103(1-3)** 2555
[23] Purcell S, Vincent P, Journet C and Binh V 2002 *Phys. Rev. Lett.* **88** **10**
[24] Wu J, Cheng Z, Sun L, Labunov V and Shulitski B 2017 *Phys. Chem. Appl. Nanostruct. (World Scientific: Singapore)*
[25] Bocharov G S and Yeletsky A V 2007 *J. Tech. Phys.* **77(4)** 107

IOP Conf. Series: Materials Science and Engineering **475** (2019) 012018 doi:10.1088/1757-899X/475/1/012018

Operating voltage diminution of field emission cathodes based on carbon nanotubes

L Sun[1], Z Cheng[1], Z Y Li[2], V Labunov[3], B Shulitski[3], I Kashko[3] and D Grapov[3]

[1]Institute of Electron Device & Application, Hangzhou Dianzi University, Hangzhou, China
[2]North-China Electromagnetic Protection Research Institute, Taiyuan, China
[3]Belarusian State university of Informatics and Radioelectronics, Minsk, Belarus

Corresponding author's e-mail address: labunov@bsuir.by

Abstract. The possibility of minimizing the operating voltages of the diode structure with a field emission cathode based on an array of carbon nanotubes is investigated. The numerical simulation method was used to perform theoretical calculations that showed the potential for an unlimited reduction in the operating voltage by reducing the diameter of the emitters, increasing the height of the emitters and reducing the value of the interelectrode gap. Established deviation of the experimental values of the operating voltage from theoretically calculated when approaching in the interelectrode gap near to values commensurate with the height of the emitters. It is shown that the observed effect is caused by low conductivity of multi-walled carbon nanotubes, high contact resistance with a substrate, as well as micro- and macroinhomogeneities of the structure of an array of carbon nanotubes.

1. Introduction

One of the promising and intensively studied fields of carbon nanotubes (CNTs) application is field-emission cathodes (FECs) [1]. FECs can be used in a series of devices, such as compact high-frequency electron sources, powerful microwave amplifiers, compact X-ray sources, sources of synchrotron radiation, high brightness flat-panel displays, high-performance luminescent lamps etc. The attractiveness of CNTs for these applications is due to their electrical, chemical, thermal and mechanical properties, that are reached owing to quasi one-dimensional structure and naturally high aspect ratio. Regardless of the field of FECs application, a number of requirements to their characteristics must be satisfied. Among them are the high emission efficiency, high homogeneity, density and stability of electron emission. Of particular interest is the field of vacuum integrated micro- and nanoelectronics – a new class of miniature vacuum electronic devices and vacuum integrated circuits based on CNT FEC [2]. These new devices have an ultra-high speed (subpicosecond), high radiation resistance, low sensitivity to temperature and a large coefficient of efficiency.

One of the key problems of vacuum integrated micro- and nanoelectronics is the problem of minimizing the operating voltage at a fixed value of the field emission current. It is associated with the limited values of the breakdown voltages of dielectrics used in microelectronics and the relatively small values of their working thicknesses. According to the existing theoretical ideas, there are no reasons preventing the reduction of operating voltages CNT FEC to arbitrarily small values by optimizing the geometry of the cathodes. In particular, an increase in the emitter amplification factor, a decrease in the diameter of the emitter and the interelectrode gap of the anode-cathode should lead to a decrease in the operating voltage to an arbitrarily small value. However, the results of experimental studies indicate that there is an effect limiting the possibility of minimizing the operating voltage of CNT FEC.

2. Experimental

The arrays of CNTs were synthesized by CVD method realized by a high temperature pyrolysis of fluid hydrocarbon (*p*-xylene [C_8H_{10}]) mixed with the volatile source of catalyst (ferrocene [$Fe(C_5H_5)_2$]) using pre-structured localized catalyst on the *n*-type silicon wafers with high doping level KES-0, 01 (100) of the size of 5×5 mm^2. The content of ferrocene in this "feeding solution" was 0,1% wt %. Process was conducted under the atmospheric pressure by using Ar as a gas-carrier. The aerosol of the feeding solution is delivered into the synthesis zone by dosed injection at a rate of $0{,}15 \div 0{,}5$ cm$^3\cdot$min^{-1}, a speed of the flow of Ar was 100 cm$^3\cdot$min^{-1}. The synthesis process was realized in the tubular type quartz reactor of the specially constructed equipment at relatively low working temperature of 650 °C and synthesis time of t = 4 minutes. The speed of the cooling was determined by the process of natural cooling of reactor.

The localized catalyst is a metallic layer of Al\Ni with thickness of 10 nm and 1 nm, respectively. To prevent interaction of Al with Si the Si-substrate was coated with molybdenum film of 200 nm thickness. The topology of the test structure is obtained by lift-off photolithography.

As it is shown in our patent [3], the use of volatile catalyst Fe from $(C_5H_5)_2Fe$ source in combination with the multilayer localized catalyst Mo\Al\Ni, playing role of the activator of metal-organic compound dissociation, insures obtaining the few-wall CNTs with narrow diameter distribution of low defect density, high adhesion to the substrate and minimum micro- and macro nonuniformity of CNTs height.

Study of the geometric features of the experimental samples of CNT FECs were carried out using a scanning (JEOL 6510) and transmission (JEOL 100CX) electron microscopes. The field emission (FE) properties of the experimental samples of CNT cathodes were investigated in DC mode using a diode-type cell with a fixed spacer (BSUIR).

The matrix structure of CNT PEC of densely packed single elements with a 5 μm step is shown in figure 1 (a, b) (SEM). The internal structure of carbon nanotubes is shown in figure 1 (c) (transmission electron microscopy, TEM). We can see a carbon nanotubes with $5 \div 10$ nm in diameter.

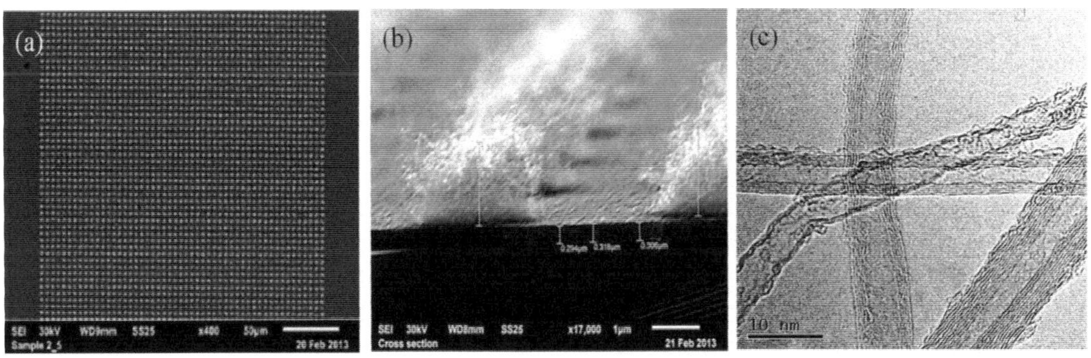

Figure 1. Microphotographs (a, b) of scanning electron microscopy of an array of carbon nanotubes; (c) a micrograph of the internal structure of carbon nanotubes (transmission electron microscopy).

3. Results and discussion

3.1 The relationship of the operating voltage and the interelectrode gap CNT FEC

The field emission current of the emitter on the basis of a single carbon nanotube is determined by the value of the local electric field intensity E_{FE} at its vertex in accordance with the classical Fowler-Nordheim formula [4-6]:

$$I_{FE} = \frac{ASE_0^2}{\varphi} \exp\left[\frac{-B\varphi^{3/2}}{E_0}\right] \text{ A/cm}^2, \tag{1}$$

where E_0 – local electric field intensity at the emitter tip (V/cm), S – emission area (cm^2), φ – work function for emitter material (eV), A = 154 and B = 6830 – constant values. For multi-walled carbon CNT φ = 5 eV [7, 8]. In its turn $E_0 = \beta \cdot E_{FE}$ – where E_{FE} – the intensity of the electric field in the gap between the tip of the emitter and the anode, β – field amplification factor.

The fixed current value uniquely corresponds to a fixed local electric field intensity E_0 at the vertex of emitter. Then, the operating voltage U_{FE} at the interelectrode gap cathode-anode required to create such an autoemission current will be determined by expression:

$$U_{FE} = \frac{E_0 \cdot (L-h)}{\beta} \tag{2}$$

where E_0 – local electric field intensity at the vertex of emitter, L – interelectrode gap between cathode and diode cell anode, β - field amplification factor near the tip of the emitter, h – emitter height.

We use the expression for the coupling of the amplification factor β and geometric parameters of the emitter on the basis of a single carbon nanotube:

$$\beta = \frac{h \cdot (1 + \frac{d}{L-h})}{d} \tag{3}$$

where h and d are respectively, the height and diameter of the nanotube, L – cathode-anode interelectrode gap.

As a consequence, we obtain the dependence of the operating voltage U_{FE} from geometric parameters h, d and L diode cell cathode-anode with a single nanotube at a fixed value of the field emission current:

$$U_{FE}(h, d, L) = \frac{E_0 L d}{h(1 + \frac{d}{L-h})} \tag{4}$$

A practical variant of the field emission cathode based on carbon nanotubes is the array of carbon nanotubes, formed on the surface of a functional substrate in the CVD synthesis. A feature of this array is a large spread of geometric parameters of carbon nanotubes in height and diameter, and the stochastic distribution of carbon emitters over the substrate surface. As a consequence, in the case of an array of carbon nanotubes as additional descriptive parameters a distance parameter appears between the individual nanotubes in the array and the parameters, characterizing the dispersion (variance) of the values of individual parameters of CNT in terms of diameter, height and distance between them.

The formula (4) allows to estimate qualitatively the character of dependence $U_{FE} = f(h,d,L)$. More detailed modeling of the dependence $U_{FE} = f(h,d,L)$ for an array of carbon nanotubes requires taking into account the effect of electron tunneling through a potential barrier near the surface of a solid in a vacuum. We carried out this simulation on the basis of a numerical solution of the Poisson equations together with the quantum-wave equation in the one-electron approximation. The calculation procedure is outlined in the work [9]. The dependences of the working voltage of the diode structure based on CNT FECs on the cathode-anode interelectrode gap at a fixed emission current density and variations in the diameter and height of carbon emitters are calculated. For each set of variables h, d and L an optimization was performed on the distance between the emitters in the array and the minimum value of the operating voltage. It was find out that the characteristic qualitative form of the obtained dependences $U_{FE} = f(h,d,D)$ practically does not depend on the density of the emission current. The results of calculations for a fixed emission current density of 1 A/cm^2 are shown in figure 2. As we can see, theoretical calculations by the numerical simulation testify to the potential for an unlimited reduction in the operating voltage due to a decrease in the emitter diameter, an increase in the emitter height, and a decrease in the interelectrode gap.

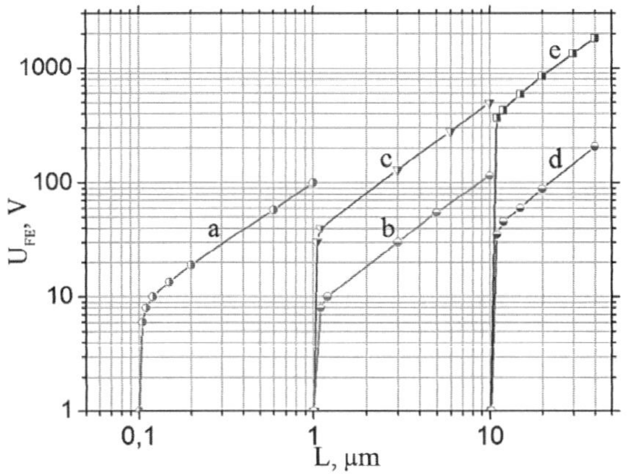

Figure 2. Theoretical dependences of the operating voltage of the diode structure based on CNT FECs on the cathode-anode interelectrode gap at a fixed emission current density of 1 A/cm^2 and variations in the diameter and height of carbon emitters: (a) – d = 1,2 nm, h = 100 nm; (b) – d = 1,2 nm, h = 1 μm; (c) – d = 12 nm, h = 1 μm; (d) – d = 12 nm, h = 10 μm; (e) – d = 120 nm, h = 10 μm.

To verify theoretical calculations, we fabricated diode structures with FECs based on arrays of carbon nanotubes and investigated the dependence of the operating voltage on the value of the interelectrode gap *L* cathode-anode for three heights of CNT – *h* = 2, 5 and 10 μm. The CNT FECs and the internal structure of synthesized emitters are shown in figure 1 (a-c).

The current-voltage characteristic of the diode structure was measured in DC mode. On the basis of the measured current-voltage characteristics, the dependence of the operating voltage on the value of the interelectrode gap is constructed for a fixed value of the emission current density of 1 mA/cm^2 (figure 3).

The dashed lines in figure 3 show the course of the theoretically calculated dependences $U_{FE} = f(L)$ at fixed values *d* and *h*. Theoretically calculated dependences are combined with those obtained by experimental measurements of the current-voltage characteristics. As we see, with a decrease in the interelectrode gap of the cathode anode to values commensurate with the height of the emitters, the deviation of the experimental curves $U_{FE} = f(L)$ from theoretically calculated is observed. Namely, with a decrease in the interelectrode gap in the experimental diode structure, the decrease in the operating voltage for a fixed value of the field emission current is rapidly slowed down.

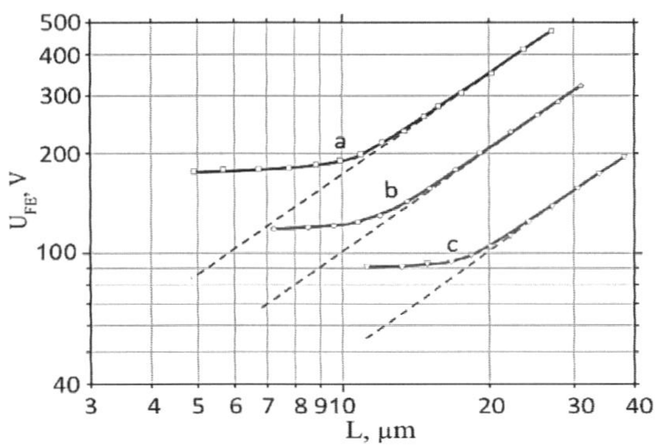

Figure 3. Experimental dependences of the operating voltage of a diode structure based on CNT FECs on the cathode-anode interelectrode gap at a fixed emission current density 1 mA/cm^2: (a) – h = 2 μm;

The registered effect is the main obstacle for the use of CNT arrays as field-emission cathodes in the field of a new perspective direction of work - vacuum micro- and nanoelectronics, since it does not allow to perform the required micro- and nano-miniaturization of field emission cathodes. In this regard, relevant work is to explain and search for possible causes of the observed effect.

3.2 Minimum operating voltage limiting mechanism of CNT FEC

There are reasons to believe that three known facts can be related to the underlying causes of the observed effect – low conductivity of multi-walled carbon nanotubes, high contact resistance with a substrate, as well as micro- and macroinhomogeneities of the structure of the CNT array.

In practice, the individual emitters, as a rule, are multi-walled CNTs. They are characterized by a significant defect structure of graphene walls, causing a low electrical conductivity of CNTs. Another problem is poor mechanical, electrical and thermal contact with the substrate. As follows from [10], the conductivity of multiwall CNTs in practice corresponds to the conductivity of graphite, therefore the resistance value of the CNT array is of the order of several megoOhms. Moreover, one should keep in mind that due to the significant dispersion of CNTs geometric parameters in the cathode (height, diameter and distance) the main contribution to the emission current yield gives at best a few percent of the CNTs. Taking into account all this together the actual resistance of the CNT emitters can reach hundreds of megoOhms.

Further, a decrease in the interelectrode gap leads to a proportional decrease in its resistance, whereas the emitter resistance and a contact resistance of emitter-substrate is constant. This situation leads to the fact that, beginning from the certain values of the interelectrode gap only some part of the applied potential falls on the discharge gap, and a significant part of the potential falls on the high resistance emitting nanotubes and their contact with the cathode substrate. As a consequence, higher operating voltages are required to maintain a fixed density of the emission current in the diode cell. Beginning with some gap size, almost all of the operating voltage applied to the electrodes of the diode structure falls on the carbon nanotubes and their contact with the cathode substrate. Thus, the minimum possible value of the operating voltage CNT FEC is determined by the resistance of the carbon emitters and their contact with the cathode substrate. It should be noted that this same mechanism underlies another known effect – the so-called emission current saturation effect [11]. This effect is manifested in the pronounced inflection of the current-voltage characteristics and Fowler-Nordheim dependences at a sufficiently large emission current densities [12].

Another reason for the observed effect of limiting the minimum operating intensity of CNT FEC may be a statistical spread of the structural parameters of individual emitters over the area of the CNT cathode, associated with the imperfection of the CVD technology of CNT synthesis. In the case of a cathode with a significant number of point emitters, an ideal homogeneity (uniformity) of the structure along the height and diameter of the individual emitters, the distance between them, and the interelectrode gap value is required. In practice, the ideal unification of the structure of CNT in processes of self-organization (growth) is technologically difficult to achieve in all known growth technologies, such as CVD, arc discharge, laser ablation, etc. In all mechanisms of synthesis, it is impossible to avoid heterogeneity in diameter, height of carbon nanotubes, and the distance between them.

As already mentioned, the emission current I_{FE} of field-emission cathodes is described by the modified Fowler-Nordheim expression (1). In accordance with the Fowler-Nordheim formula, as well as its generalization for thermofield emission, the Murphy-Hood formula [13], the values I_{FE} and E_0 are related by an exponential dependence. In accordance with the formula (2) the same exponential relationship exists between I_{FE} and U_{FE}. Depending on the CVC area, the increment U_{FE} at 10 % may lead to an increase of I_{FE} of an individual emitter in 5 ÷ 10 times. On the other hand, the statistical variation in height and diameter of individual emitters can significantly affect U_{FE}, since the height of the emitter determines the interelectrode gap between the emitter and the anode, and the emitter diameter is the internal field amplification factor β on an individual emitter. The interrelationships of U_{FE} with h, d and L follow from the formula (4).

The statistical spread of the geometric parameters of neighboring tip emitters (individual carbon nanotubes) can be defined as microinhomogeneities of the structure. At the same time, also the heterogeneity of the CNT parameters at different parts of the cathode matrix begins to play a role on a large working area of the cathode, determined by the inhomogeneity of the growth conditions of CNT on the extended synthesis zone in the CVD reactore. Such heterogeneity can be defined as macroinhomogeneity. In the process of CVD synthesis, the microinhomogeneity of the array structure (the spread of the geometric parameters of neighboring CNTs in the array), as a rule is $5 \div 10$ % from average values of height and diameter of CNT. With a decrease in the interelectrode gap, an increasing proportion of CNT with a lower coefficient of amplification of the electric field (small height and large diameters) are excluded from effective emission, since due to the exponential nature of the dependence I_{FE} on U_{FE} their emission currents drop sharply. A similar contribution is made by the macroinhomogeneity of the CNT array, determined by the heterogeneity of the growth conditions of CNT in the extended zone of synthesis of the CVD reactor. The effect is manifested in a decrease in the area of effective emission relative to the cathode working area. A decrease in the area of effective emission leads to a reduction in the total emission current. As a consequence, the preservation of the value of the emission current stationary requires higher operating voltages at the electrode gap cathode-anode, which compensates for the decrease in the area of effective emission. Negative result is overload, burning out of the most active emitters and destruction of the cathode.

Thus, the low conductivity of multi-walled carbon nanotubes, the high contact resistance with the substrate, and the micro- and macroinhomogeneities of the structure of the CNT array are the main reason for limiting the minimum operating voltage of the diode cell with CNT FEC.

4. Conclusions

The possibility of minimizing the operating voltages of field emission cathodes on the basis of arrays of carbon nanotubes is investigated. Theoretical calculations are performed by the numerical simulation method, which showed the possibility of an unlimited reduction in the operating voltage of CNT FECs by reducing the diameter of the emitter, increasing the emitter height and reducing the value of the interelectrode gap.

To test the theoretical calculations, studies of the dependence of the operating voltage CNT FECs on the value of the interelectrode gap L cathode-anode for three heights CNT – h = 2, 5 and 10 μm were carried out. CVC were measured in a diode structure in an DC voltage mode.

It is shown that when the cathode-anode inter-electrode gap is reduced to values commensurate with the height of the emitters, the deviation of the experimental curves $U_{FE} = f(L)$ from theoretically calculated is observed. In practice, with a decrease in the interelectrode gap, the decrease in the operating voltage for a fixed value of the autoemission current is rapidly slowed down.

It is shown that the observed effect is caused by low conductivity of multi-walled carbon nanotubes, high contact resistance with substrate, as well as micro- and macroinhomogeneities of the CNT array structure.

References

[1] Yeletsky A V 2002 *Phys. Usp. (Russia)* **45(4)** 369
[2] Woo H J, Sub O J and Meyyappan M 2012 *Appl. Phys. Lett.* **100** 213505
[3] Saurov A N et al 2012 *Patent of the Eurasian Patent Office* EA016557
[4] Fowler R H and Nordheim L 1928 *Proc. Roy. Soc. Lon.* **A119** 173
[5] Burgess R E, Kroemer H and Houston J M 1953 *Phys. Rev.* **90** 515
[6] Kim D, Bouree J E and Kim S Y 2006 *Appl.Phys. A* **83** 111
[7] Gröning O, Küttel O M, Emmenegger C et al 2000 *J.Vac. Sci. Technol. B* **18** 665
[8] Bonard J M, Croci M, Arfaoui I et al 2002 *Diamond Relat. Mater.* **11** 763
[9] Pozdnyakov D V, Borzdov A V, Borzdov V M 2013 *Proceeding of the international conference: Nanomaterials, applications and properties* **2(1)** 01001
[10] Kataura H, Kumazawa Y, Maniwa Y et al 1999 *Synthetic Metals* **103(1-3)** 2555
[11] Milne W I, Teo K B, Minoux E at al 2006 *J. Vac. Sci. Technol. B* **24(1)** 1074
[12] Murphy E L and Good R H 1956 *Phys. Rev.* **102** 1464
[13] Wu J, Cheng Z, Sun L, Labunov V and Shulitski B 2017 *Physics, Chemistry and Application of Nanostructures (World Scientific: Singapore)*

Europium luminescence from amorphous yttrium alumina films on fused silica substrates

J C Villegas Brito[1], N V Gaponenko[1,4], K S Sukalin[1], T F Raichenok[2], S A Tikhomorov[2], Xiang Wang[3], Zhiqun Cheng[3], N I Kargin[4]

[1]Belarusian State University of Informatics and Radioelectronics, P. Browki St. 6, 220013 Minsk, Belarus

[2]Stepanov Institute of Physics, NASB, Nezavisimosti ave. 68, 220072 Minsk, Belarus.

[3]Key Laboratory of RF Circuits and Systems, Ministry of Education, Hangzhou Dianzi University, 310018, Hangzhou, China

[4]National Research Nuclear University MEPhI (Moscow Engineering Physics Institute), Kashirskoe Shosse 31, 115409, Moscow, Russian Federation

Corresponding author's e-mail address: nik@nano.bsuir.edu.by

Abstract. In this work, we investigate photoluminescence of sol-gel derived films corresponding to yttrium alumina garnet $Y_3Al_5O_{12}$ doped with trivalent europium. The films with three concentrations of europium 0,02; 0,025; 0,03 M with the thickness 210-220 nm were deposited on fused silica substrates by the corresponding sols' spinning followed by drying and high temperature annealing at 1000 °C for 30 min. The samples were characterized by X-ray diffraction (XRD), scanning electron microscopy (SEM), energy dispersive X-ray (EDX) analysis as well as ellipsometry, photoluminescence (PL) and photoluminescence excitation spectroscopy (PLE). All of the fabricated samples were amorphous and revealed room-temperature luminescence bands within the range of 550 – 750 nm corresponding to $^5D_0 \rightarrow {}^7F_j$ (j=0,..,4) transitions of trivalent europium ions with the most intensive bands at 610 – 630 nm and 680 – 710 nm. The intensity of europium luminescence increases with europium concentration. The photoluminescence excitation spectra measured at the emission wavelength 710 nm exhibit strong bands at 220 nm, which could be associated with absorption of exciting light within Eu_3^+ - O_2^-bands. Practical application of the obtained films transforming ultraviolet irradiation into visible is discussed.

1. Introduction

Materials that are transparent in the visible and near infrared range when doped with light emitting impurities receive interest as light convertors of UV irradiation into visible, which increases the efficiency of silicon solar cells [1]. Particularly, trivalent europium ions are characterized by strong visible luminescence [2]. Luminescence properties of Eu-doped materials (particularly of the PL and PLE spectra) depend on host material, synthesis conditions and annealing temperature, as well as europium concentration. For most of the host materials, the band at 610 nm is the most intensive as compared to the band at 710 nm. However, for silicon-based photodetectors, conversion of UV irradiation into emission at about 710 nm is of great importance, considering their spectral sensitivity. Recently, strong Eu PL with efficient 710 nm emission was observed from aluminates codoped with Eu and Bi [3]. Eu-doped film structures, which exhibited strong visible PL, were created using magnetron sputtering, laser ablation, molecular beam epitaxy, deposition from the salt solutions and other techniques. The properties of sol-gel derived films depend on the type of substrate. In this work we characterized Eu-doped amorphous yttrium alumina films with europium concentrations of 0,02, 0,025, and 0,03 M deposited on fused silica substrates.

2. Experiment

The components of the sol preparation were ethanol (C_2H_5OH), citric acid ($C_6H_8O_7$), aluminium nitrate nonahydrate $Al(NO_3)_3$ $9H_2O$, 98.5% (Riedel-de Haen), pentahydrate nitrate, europium (Eu $(NO_3)_3$ $5H_2O$, 99.9%, Aldrich) and yttrium nitrate pentahydrate ($Y(NO_3)_3 \cdot 6H_2O$, 99.999%, Aldrich). Preparation of sols with different europium concentrations was carried out by changing the mixing ratio of europium and yttrium, which ensured the concentration of Eu^{3+} in three sols 0.02; 0.025 and 0.03 M. Sols were deposited on a quartz substrate by spinning at a rate of 2700 rpm for 30 seconds. After drying at 200 °C of the first layer, the second layer was deposited and dried at the same conditions. Finally, the samples were subjected to heat treatment at 1000 °C for 30 min.

The photoluminescence (PL) and PLE spectra were studied using the spectrofluorometer CM2203, which allows measurements of luminescence and excitation spectra in the range of 210-800 nm. A high pressure xenon arc lamp with 150 Watts was used as an excitation source. In the channels of excitation and registration of luminescence, double monochromators with dispersion summation were installed. To reduce the penetration of ambient light so as to reduce the influence of scattered light, the studied samples were mounted in a special holder of the detection channel at an angle of 30° between the normal to the sample plane and exciting beam.

X-ray diffraction analysis was performed on the diffractometer D8 Advance, Bruker AXS (Germany), using CuK α-radiation. Morphological analysis of the films was investigated using a HITACHI S-4800 scanning electron microscope equipped with an X-ray microanalyzer for EDX-analyses. The refractive index and the film thickness were measured by laser ellipsometry 3M LEF-1 with a wavelength of 6328 Å.

3. Results and discussion

The EDX analyses performed for the sample with concentration 0,025 M confirms that the yttrium alumina composite contains the elements in the ratio (Y+Eu)/Al=3:5 with the accuracy of experimental error. The EDX analyses performed for the sample with concentration 0,025 M confirms within range of the accuracy of the experiment of that the yttrium alumina composite contains the elements in the ratio (Y+Eu)/Al=3:5. The XRD spectra revealed that the films doped with three Eu concentrations are amorphous (Figure 1). According to SEM-data, the typical thickness of the films is about 210 nm and vague images of grains are slightly resolved. (Figure 1 in the inset). For the film of 216 nm thickness, the refractive index was found to be 1.81, according to ellipsometry analysis.

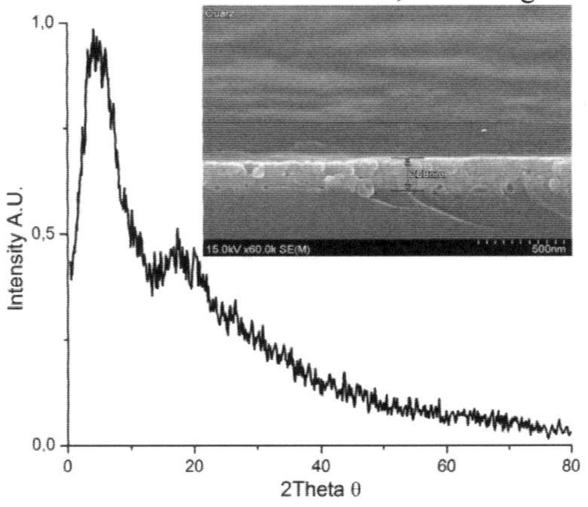

Figure 1. XRD spectra of Eu-doped yttrium alumina film on fused silica substrate after annealing at 1000 °C for 30 min for Eu concentrations 0,03 M. XRD spectra for 0,02 and 0,025 M is basically the same. SEM-image of the film with Eu concentration 0,02 M is given in the inset.

Figure 2 gives the PL spectra of these films with three Eu concentrations recorded for the excitation wavelength 260 nm. For all of the fabricated films, the PL spectra reveal the bands corresponding to optical transitions between the terms of trivalent europium ions: $^5D_0 \rightarrow {}^7F_1$ (585 - 600) nm; $^5D_0 \rightarrow {}^7F_2$ (610 - 630) nm; $D_0 \rightarrow {}^7F_3$ (640 - 660) nm; 5 $^5D_0 \rightarrow {}^7F_4$ (680 – 710) nm [4,5]. The PL intensity of all of the observed bands increases with Eu concentration. For the sample with the highest Eu concentration, additional weak band 570 - 585 nm is observed corresponding to $^5D_0 \rightarrow {}^5F_0$ transition.

PLE spectra of Eu-doped films with three concentrations 0,02; 0,025 and 0,03 M were investigated for the most intensive 710 nm $^5D_0 \rightarrow {}^7F_4$ band. The observed bands in PLE spectra correspond to trivalent Eu transitions between the terms: $^7F_0 \rightarrow {}^5H_6$ (318 nm); $^7F_0 \rightarrow {}^5D_4$ (365 nm); $^7F_0 \rightarrow {}^5G_2$ (380 nm); $^7F_0 \rightarrow {}^5L_6$ (393 nm); $^7F_0 \rightarrow {}^5D_3$ (413 nm)[6]. The intensive wide band 214 - 280 nm is associated with the absorption of charge transfer band Eu^{3+} - O^{2-} [7,8].

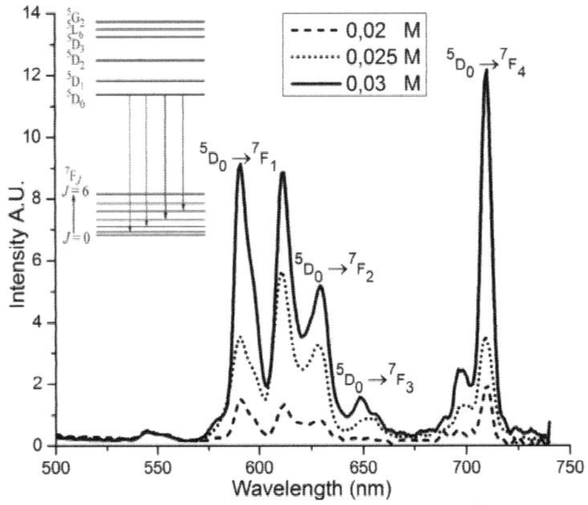

Figure 2. PL spectra of Eu-doped yttrium alumina films obtained for 3 concentrations of europium after annealing at 1000 °C for 30 min: (Eu^{3+}) 0.02, 0.025, 0.03 M. Excitation wavelength λ_{ex} = 260 nm.

Transmission spectra of yttrium alumina composites with Eu concentration 0.03 M are presented in Figure 4. The absorption bands in the range 200—394 nm correspond to the bands in PLE spectra.

Eu-doped films could find application as transparent convertors of UV-irradiation into visible [9], raising the efficiency of silicon solar cells and converting UV irradiation into visible due to reducing electron thermalization and degradation of the solar cell. Eu-doped amorphous films also receive interest in the investigation of spontaneous emission properties when embedded in microcavities [10], photonic crystals [11], lighting sources [12] and other matrices with anisotropy of photonic density of states [13, 14].

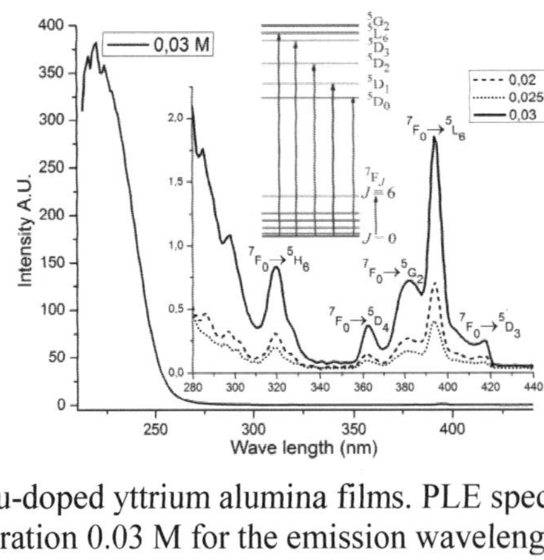

Figure 3. PLE spectra of Eu-doped yttrium alumina films. PLE spectrum in the range 214-440 nm of Eu-doped film with concentration 0.03 M for the emission wavelength 710 nm. The inset depicts PLE spectra in the range 280 – 440 nm for the samples with three europium concentrations: 0.02, 0.025 and 0.03 M.

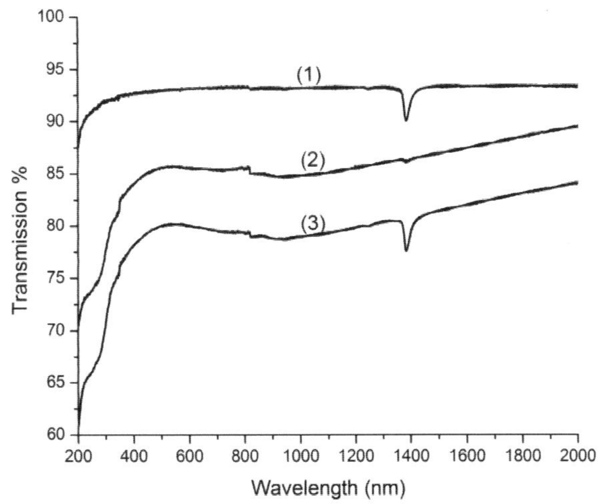

Figure 4. Transmission spectra: (1) – uncoated fused silica substrate; (2) - fused silica substrate with yttrium alumina film with Eu concentration 0.03 M in reference channel, (3) fused silica substrate with yttrium alumina film with Eu concentration 0.03 M without fused silica substrate in the reference channel.

4. Conclusion

The method of sol-gel synthesis of Eu-doped yttrium alumina films has been described in this article. The films, comprising two layers deposited by spinning on fused silica substrate, are about 210 nm thick and are amorphous after conventional furnace annealing at 1000 °C. The films reveal visible Eu-related PL with the bands $^5D_0 \rightarrow {}^7F_1$ (588 nm), $^5D_0 \rightarrow {}^7F_2$ (608 and 629 nm), $^5D_0 \rightarrow {}^7F_3$ (654 nm), $^5D_0 \rightarrow {}^7F_4$ (710 nm) excited with UV irradiation in the range of 200-394 nm. The PL intensity increases for the chosen concentrations of Eu – 0,02; 0,025 and 0,03 M. The described method allows the strong conversion of UV irradiation into visible with the most intensive 710 nm PL band. The obtained film has an amorphous structure after a heat treatment at 1000 ° C.

Acknowledgements

The authors wish to thank Dr. G.E. Malashkevich for useful discussion, S.V. Husakova for the performed EDX analyses and D.V. Zhugulin for SEM analyses.

References

[1] Van Sark W G J H M, Meijerink A and Schropp R E I 2012 *Third Generation Photovoltaics (Netherland: Dr. Vasilis Fthenakis)* 232
[2] Binnemans K 2015 *Coordination Chem. Rev.* **295** 1
[3] Malashkevich G E, Shevchenko G P, Bokshitz Yu B, Kornienko A A and Pershukevich P P 2015 *Opt. Spec*. **98** 190
[4] Sandro M L, ALuiset H da C and Junior R S 2012 *Opt. Express*. **20** 12658
[5] Gaponenko N V, Molchan I S and Thompson G E 2002 *Sensors Actuators* **99** 71
[6] Ling A, Li Zh and Liu J 2015 *2nd Int. Conf. Modeling Identification Control*
[7] Gaponenko N V, Molchan I S and Gaponenko S V 2003 *J. Appl. Spectrosc*. **70** 57
[8] Esparza A, Garcia M and Falcony C 1998 *Thin Solid Film*. **325** 14
[9] Potdevin A, Lechevallier S and Chadeyron G 2009 *Opt. Mat.* **517** 4610
[10] Bellessa J, Rabaste S and Plenet J C 2001 *Appl. Phys. Lett.* **79** 14
[11] Potdevin A, Chadeyron G and Boyer D 2005 *J. Phys. D: Appl. Phys.* **38** 3251
[12] Petrov E P, Bogomolov V N, Kalosha I I and Gaponenko S V 1998 *Phys. Rev. Lett.* **81** 77
[13] Lutich A, Gaponenko S V and Gaponenko N V 2004 *Nano Lett*. **4**1 755
[14] Gaponenko N V 2007 *Acta Phys. Pol.* **112** 737

IOP Conf. Series: Materials Science and Engineering **475** (2019) 012020 doi:10.1088/1757-899X/475/1/012020

Design and fabrication of terahertz quantum cascade laser with double metal waveguide based on multilayer GaAs/AlGaAs heterostructures

R A Khabibullin[1], N V Shchavruk[1], A Yu Pavlov[1], A N Klochkov[1], I A Glinskiy[1], K N Tomosh[1], D S Ponomarev[1], G E Cirlin[2] and A E Zhukov[2]

[1]Mokerov Institute of ultra high frequency semiconductor electronics of RAS, Moscow, Russia
[2]St Petersburg National Research Academic University of RAS, St Petersburg, Russia

Corresponding author's e-mail address: khabibullin@isvch.ru

Abstract. We have designed multilayer heterostructure (MH) based on three quantum well GaAs/Al0.15Ga0.85As active module with diagonal transitions and optimized oscillator strength – 0.425. Furthermore, we have developed a method for the fabrication of terahertz quantum cascade laser (THz QCL) with double metal waveguide via low-temperature In-Au wafer bonding followed by SI-GaAs substrate removal. ICP RIE in BCl3/Ar has been used to obtain laser ridges (widths 50-200 mkm) with vertical sidewalls. We investigate the dependence of the MH energy band structure on the electric field and thermal properties of terahertz quantum cascade lasers (THz QCL) under different operation conditions.

1. Introduction

The development of technology for manufacturing of devices operating in THz range can give the global medical science new principles and theoretical approaches in early diagnosis and healing of chronic diseases (especially tumors). Relevance of these developments are also linked to the existence of the problem of countering terrorism in public places and transport because there is pentup demand reliable, but also safe for human inspection systems.

Among various techniques for THz generation [1-3], THz quantum cascade lasers (THz QCL) are compact solid-state source with electrical pump. Thus, THz QCL are considered to be the most promising THz source. THz QCL based on AlGaAs/GaAs heterostructures operate in the range 1.2-5.4 THz with output powers of more than 100 mW in continuous wave regime [4] and a peak powers of more than 1 W in the pulsed mode [5]. The spectral width of a THz QCL with distributed-feedback (DFB) is equal to ~10 kHz, it allows the use of THz QCL as a local oscillator for heterodyne detection [6]. Frequency comb based on THz QCL with a spectral bandwidth of more than 1 THz was demonstrated, which enables the development of THz spectrometers with a large S/N ratio [7,8]. One of the key desired THz QCL characteristic is RT operation, but it has not been achieved yet.

In the present article we have proposed the design of GaAs/$Al_{0.15}Ga_{0.85}As$ multilayer heterostructure (MH) with optimized oscillator strength and investigated thermal properties and energy band profile of THz QCL under different operation conditions.

2. Results and discussion

On the basis of the calculation of the oscillator strength for the transitions between quantum-well levels two MH based on the three coupled quantum well (3QW) with resonant-phonon depopulation scheme (RP) were grown by molecular beam epitaxy (MBE). The layer sequences, starting from the injector barrier, are 43/75.6/24.6/69.3/41/136 (MH-1); 43/89/24.6/81.5/41/160 (MH-2) in Åangstrom. In Ref [9] the results of X-ray diffraction and photoluminescence spectroscopy characterization of the grown MH are presented.

3QW MH with four electron levels (see figure 1) is investigated. Levels E_2 and E_3 in the double quantum well are active laser states. Difference $E_3 - E_2$ corresponds to the energy and frequency of the emitted THz radiation. Levels E_1 and E_4 in the wide QW correspond to the injector and extractor levels, respectively. The THz QCL generation is possible for the following level alignments: (1) injector E_1 is aligned with higher laser state E_3, (2) lower laser state E_2 is aligned with extractor state $E_{4'}$ in the next period. The heterostructure is designed in such a way that difference $E_4 - E_1$ is equal to optical phonon energy (E_{LO} = 32 meV for GaAs) to provide resonant electron extraction.

Figure 2 shows the first level alignment of the lower injector state (E_1) with the upper injector state in the next stage ($E_{4'}$) occurs under 8.5 kV/cm (calculated within a Schrodinger-Poisson approach [10,11]). In that case the electrons tunnel through "lasing double-well" without emission of THz photons. The second level alignment occurs under 12.5 kV/cm and indicates alignment of the lower injector state (E_1) with the upper laser level (E_3). At higher bias THz QCL band structure is totally misaligned.

To estimate the thermal properties of THz QCL under the different operation conditions (power and cooling modes) modeling of heat flow processes in these devices was carried out [12]. The results for continuous wave regime for THz QCL are shown in figure 3. Due to the high value of thermal conductivity of the MH and In–Au bonding layer, thermal resistance of the device is dominated by the temperature drop inside the active region, and spreading resistance in the substrate.

The thermal dynamics of a THz QCL operating in pulse mode is investigated. In figure 4 maximum values of active region temperature T_{AR} under the different operation conditions (pulse repetition frequencies and duty cycles) are shown. Temperature-time profile for a standard ridge waveguide (inset figure 4) showing that T_{max} is stabilized for a time of about 2 ms and the further operation of the device remains unchanged.

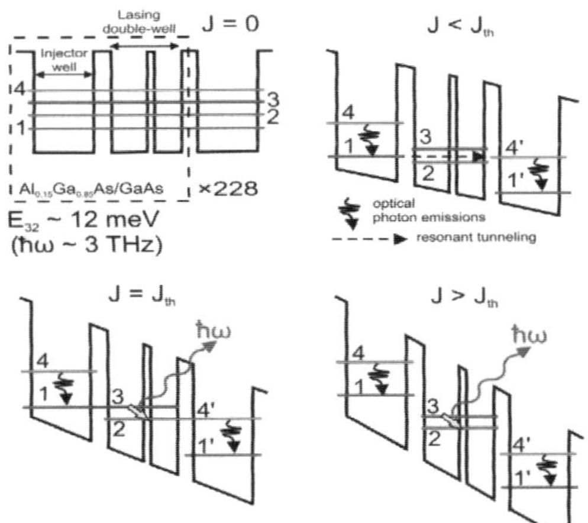

Figure 1. Schematic diagram of conduction-band of the 3QW MH based on RP scheme at different biases. A single quantum cascade stage is marked by a box. The radiative transition and LO photon emission is from 3 to 2 and from 4 to 1, respectively.

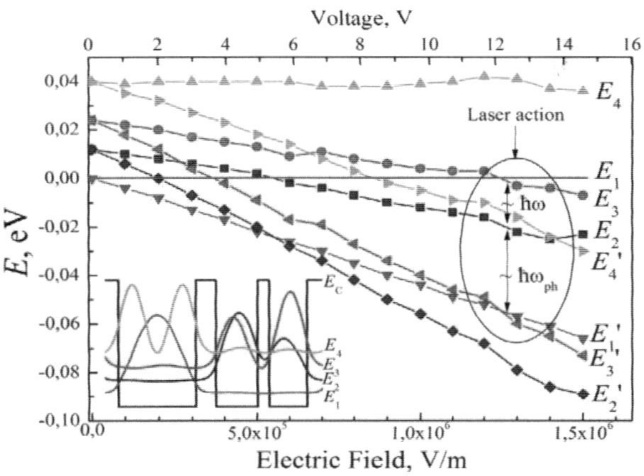

Figure 2. Dependence of the energy level positions in the current E_i and following $E_{i'}$ THz QCL periods on electric field for structure MH-2 (level E_1 is reference); inset – conduction band scheme, electron wave functions and energy levels alignment at zero electric field.

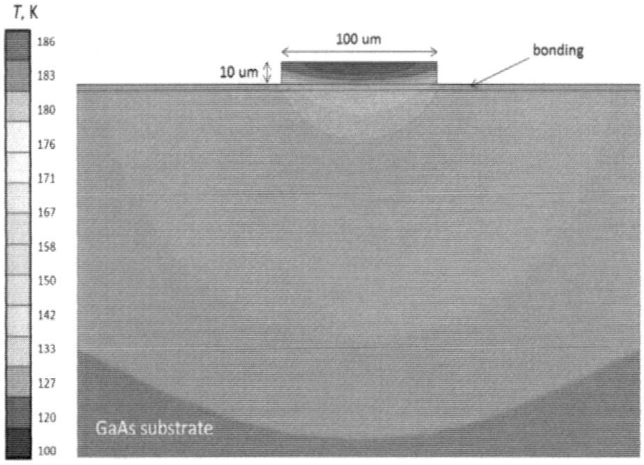

Figure 3. 2D heat flow model of cw THz QCL calculated with finite-element solver. The lower boundary is set to 100 K.

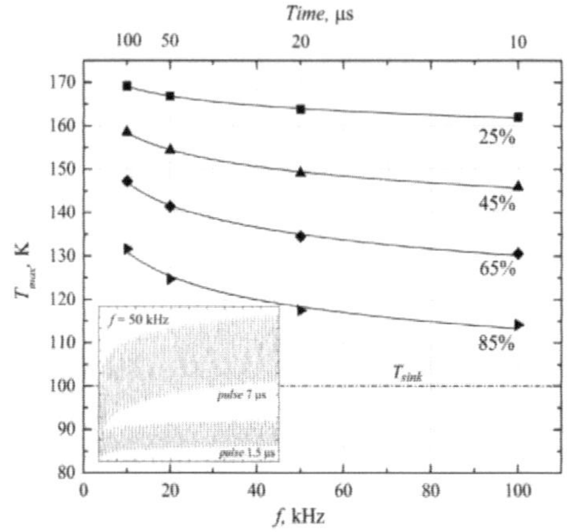

Figure 4. Maximum values of T_{AR} for different pulse repetition frequencies in the range 10-100 kHz and duty cycle 25-85%; inset – temperature-time profile for 1.5 and 7 μs pulse widths at 50 kHz.

We have developed post-growth processing of multilayer heterostructures GaAs/AlGaAs THz QCL with In–Au double metal waveguide [13]. The post-growth processing includes In–Au bonding of the THz QCL on n+-GaAs wafer, mechanical lapping and selective wet etching of the SI-GaAs wafer, and dry etching of the ridge structures using 50-μm, 100-μm and 200-μm-wide Ti/Au contacts as self aligned etch masks. The inductively coupled plasma reactive ion etching regime has been optimized in BCl_3/Ar for vertical sidewalls of the laser ridge with the minimum of Ti/Au mask destruction. Device processing then continued according to the standard recipe.

THz QCL bar is soldered onto Cu blocks, which serves as a heat sink and bottom electrical contact (figure 5). Top QCL contact is made by wire bonding to Ti/Au pads. A scanning electron micrograph of a typical wire bonding 50-μm-wide ridge structure is shown in figure 6. It has been found that n+-GaAs receptor must be thinned to ~100 μm to obtain a good quality of laser facet.
Light-current curves and spectral characteristics of fabricated lasers are presented in [14,15].

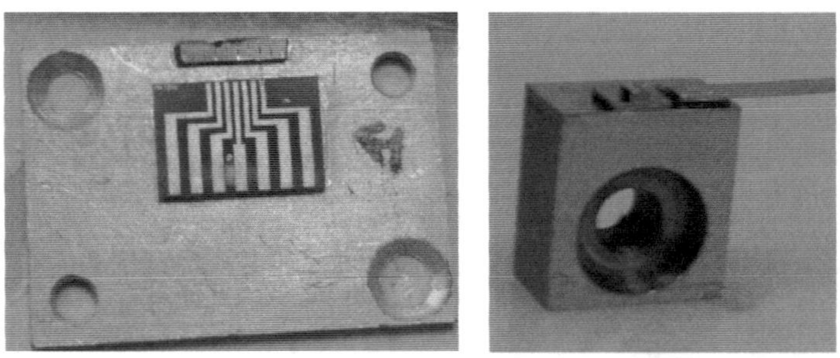

Figure 5. THz QCL bar is mounted onto C-mount (Cu blocks).

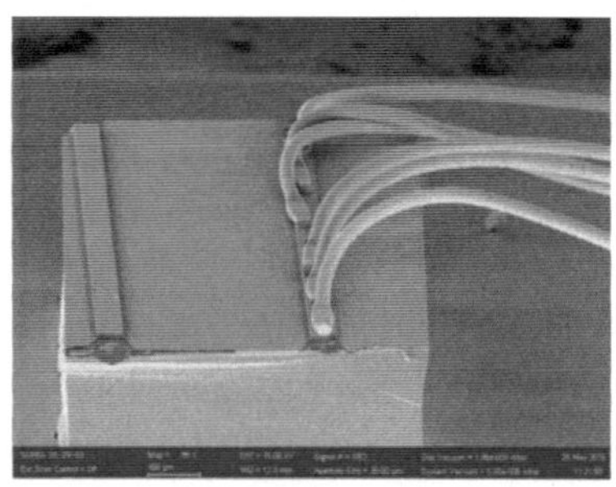

Figure 6. SEM photograph of a processed 100-μm-wide laser ridge with wire bondings.

3. Conclusion

We have proposed the multilayer heterostructure based on three quantum well GaAs/Al0.15Ga0.85As with optimized oscillator strength – 0.425. Energy band calculations and thermal modelling of THz QCL under different operation conditions were carried out. THz QCL with 50 and 100 -μm-wide ridge structure were fabricated.

Acknowledgements
This work was supported by the Russian Science Foundation (RSF), Project # 18-19-00493.

References

[1] Gergel V A, Gorshkova N M, Khabibullin R A et al 2018 *Proceedings of the Scientific-Practical Conference "Research and Development - 2016"* 101
[2] Khabibullin R A, Morozov O G, Sakhabutdinov A J et al 2018 *Systems of Signals Generating and Processing in the Field of on Board Communications*
[3] Sakhabutdinov A J, Morozov O G, Nureev I I et al2018 *Proc. of SPIE - The International Society for Optical Engineering* 10774
[4] Wang X, Shen C, Jiang T et al 2016 *AIP Adv.* **6(7)** 075210
[5] Li L H, Chen L, Zhu J X, Freeeman J et al 2014 *Electron. Lett.* **50** 309
[6] Qin Q, Reno J L, Hu Q 2011 *Opt. Lett.* **36(5)** 692
[7] Roben B, Lu X, Hempel M, Biermann K et al 2017 *Opt. Express* **25(14)** 16282
[8] Li H, Laffaille P, Gacemi D, Apfel M et al 2015 *Opt. Express* **23(26)** 33270
[9] Reznik R R, Kryzhanovskaya N V, Zubov F I et al 2017 *J. of Phys.: Conf. Series.* **917** 052012
[10] Khabibullin R A, Shchavruk N V, Pavlov A Y et al 2016 *International Journal of High Speed Electronics and Systems* **25(03-04)** 1640022
[11] Khabibullin R A , Shchavruk N V, Klochkov A N et al 2017 *Semicond.* **51(4)** 514
[12] Khabibullin R A, Shchavruk N V, Ponomarev D S 2018 *Semicond.* **52(11)** 1381
[13] Khabibullin R A, Shchvruk N V, Pavlov A Yu 2016 *Semicond.* **50(10)** 1377
[14] Ikonnikov A V, Marem'yanin K V, Morozov S V 2017 *Tech. Phys. Lett.* **43(4)** 358
[15] Volkov O Yu, Dyuzhikov I N, Logunov M V *J.Commun. Tech.Electr.* **63(9)** 1042

The Formation of a Height-Modulated Dielectric Film

A O Zotov[1], A A Perevalov[1] and I V Sagunova[1]

[1]National Research University of Electronic Technology Russia, 124498, Moscow, Zelenograd, Shokina Square, Bld. 1

Corresponding author's e-mail address: sagunova@list.ru

Abstract. The local anodic oxidation process allows creating a constantly-heighted local dielectric mask. In the course of mask formation the scanning probe moves over a particular surface area applying simultaneous constant bias voltage in each scanning point (these bias voltage impulses have the same value and duration). It is obvious that if we need to create a dielectric mask with a different oxide thickness or a height-modulated film, it is necessary to vary the impulse duration or its value at different scanning field areas. This investigation illustrates the possibility of creating a height-modulated oxide by applying the method of local anodic oxidation based on any greyscale modulated mask when the impulse duration and its magnitude is varied proportionally to the gray scale intensity.

1. Introduction

The investigation of local probe oxidation process of solid bulk materials and thin conducting films started a long time ago [1]. In contrast to the present-day projection-type photolithography where extreme ultraviolet is currently used for structure formation, the probe nanolithography based on local anodic oxidation is much more cost-effective and available for research laboratories. This method can be efficiently used both for local modification of different hard materials' properties and for the study of various active and passive components in nanoelectronics. The theoretical underpinnings of the procedure are being constantly advanced [2].

Earlier, there were theoretical attempts to make the process of height modulated oxide film formation multifunctional [3].

In the present work it is shown the peculiarities of height modulated dielectric mask formation over solid state surfaces including wafers and thin metallic films with respect to present fabrication process limitations and intrinsic natural limitations of the oxidation in the tip to surface the chemistry.

2. Selection of optimal pressing force of the conductive cantilever to the surface

A lot of semiconducting materials and metal films consist of natural oxide with adsorption surface water [4]. Consequently, successful performance of electrical measurements requires determination of the minimal pressing force on the conductive cantilever to the researched surface, which provides a close contact with the sample [5]. Therewith, using conductive cantilever of good quality is also very important in nanolithography method. Investigating outline of a volt-ampere characteristic can clarify both the minimal pressing force on the conductive cantilever to the surface and reliability conductive coating of cantilever.

The pressing force of different conductive coatings of the cantilever to the surface was investigated on the high-alloy silicon wafer. Conductive coating thicknesses of the cantilever were: W_2C and TiO_{2-x} 5nm, Pt and Au-15nm. It should be noted that the hardness of selected materials decreases array: W_2C,

IOP Conf. Series: Materials Science and Engineering **475** (2019) 012021 doi:10.1088/1757-899X/475/1/012021

TiO_{2-x}, Pt, Au. Afterwards, tip of the cantilever landed to the surface and the pressing force was determined by the force- height spectroscopy in atomic force microscopy. Simultaneously with the detection the force- height spectroscopy the volt-ampere characteristics of the system "conductive coating of a cantilever to conductive sample" were measured for different the pressing force. The pressing force was calculated according to hardness of the lever that was given by the fabricator.

Figure 1 demonstrates volt-ampere characteristics using W_2C conductive coating cantilever with various pressing force to the high-alloy silicon surface.

For little pressing forces of a tip there are non linear volt-ampere characteristics (Figure 1, *a-c*) of the system "conductive coating of a cantilever - dielectric layer- high-alloy conductive silicon sample".

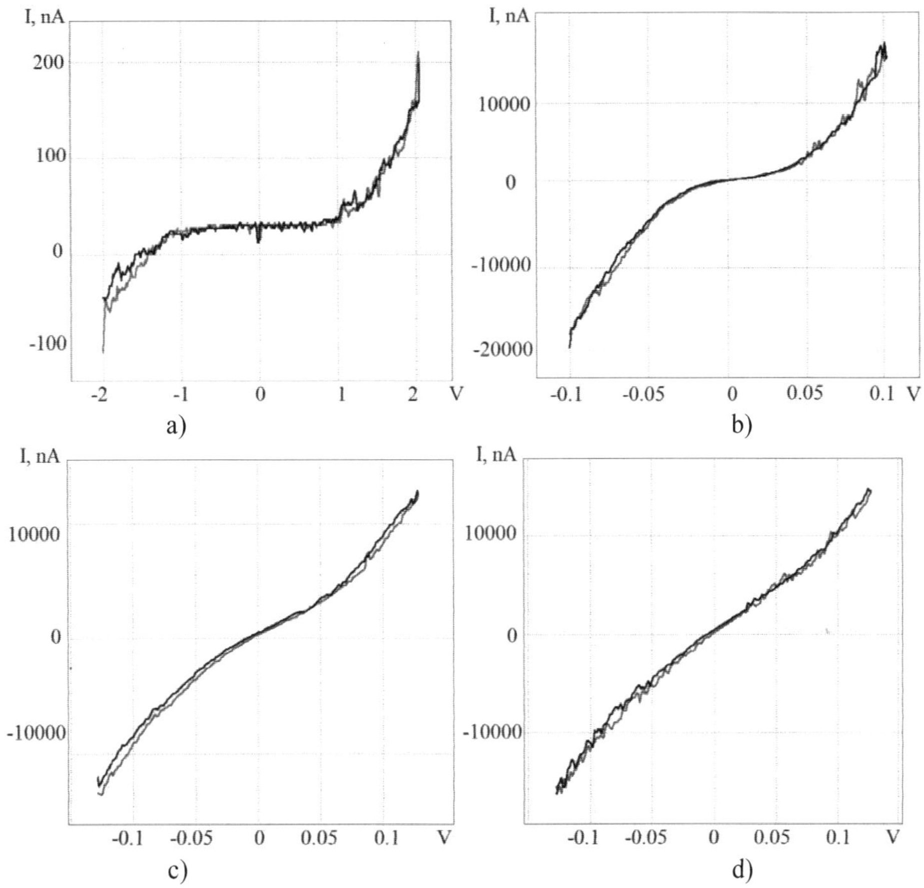

Figure 1. Volt-ampere characteristics of W_2C cantilever with various pressing force to the surface: a)-80nN, b)-150 nN, c)- 200 nN, d)- 250nN.

In beforementiones case (a) the dielectric layer is a double-layer structure: the adsorbed layer of water –natural oxide of silicon, and electrical breakdown has begun under 1V. The strong ohmic contact is achieved under 250nN pressing force and more, when we see linear outline of a volt-ampere characteristic (Figure 1, d).

We were revealed that cantilever with TiO_{2-x} coating has close ohmic contact under 280nN pressing force, with Pt coating – 350nN, with Au – 390nN. Minimum required pressing force depends on thickness of conductive coating (distinction in value of radius conductive tips) and hardness of material coating. Using sharp tips in strong ohmic contact the pressing force should be less.

IOP Conf. Series: Materials Science and Engineering **475** (2019) 012021 doi:10.1088/1757-899X/475/1/012021

Hence, for effective research the optimal pressing force of the conductive cantilever to the investigation surface should be found for specific cases depending on the material being used and conductive cantilever.

3. The process of height modulated oxide film

Formerly, there were attempts to create the constantly-heighted local dielectric mask with a different oxide thickness [6]. For instance, a mask of this sort could be used while developing highly-alloyed areas with complex doping profiles in semiconductor technology by the ion-implantation method as well as in nanoelectronic devices production, etc.

This nanolithography method is based on the following assumptions.

The local anodic oxidation process allows creating a constantly-heighted local dielectric mask. In the course of mask formation the scanning probe moves over a particular surface area applying simultaneous constant bias voltage in each scanning point (these bias voltage impulses have the same value and duration). Thus, it is obvious that if we need to create a dielectric mask with a different oxide thickness or a height-modulated film, it is necessary to vary the impulse duration or its value at different scanning field areas. It should be noted that this approach enables us to attain either step variations of the oxide height or a continuously changing oxide height. Essentially, the height modulation of the mask is determined by the function of impulse parameters in different scan areas. Nonetheless, such an easy approach is extremely effective in reproducible dielectric mask formation in accordance with the required modulation.

Specifically, the process of local dielectric height-modulated mask formation can be demonstrated experimentally by creating a nanometer size structure copy using a grey scale modulated mask of a digital photography.

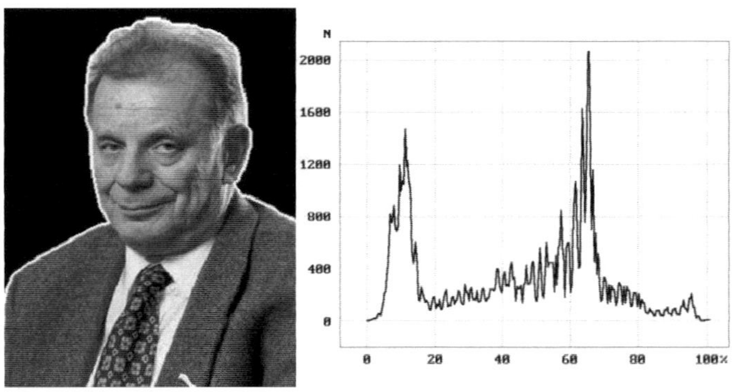

a) b)

Figure 2. The digital photography of Alferov J I a) and the gray scale gradation in percentage values b).

The process consists of the following steps. Initially, the scan pattern of the mask is uploaded into the scanning probe microscope software and projected over a particular surface area. By a scan pattern we mean a three-dimensional array of pixels each characterized by two coordinates and the grey scale intensity values expressed as percentage. Further, the titanium film deposited over a silicon wafer is scanned at amplitude modulation mode in order to get a clean area of the thin film with no defects or dust. Then, via the use of a statistical analysis, the gray scale mask is converted to percentage values in the range of 0 – 100%.

As the mask we have chosen the digital photography of Nobel prize winner in physics professor Alferov J I as the idea based on which we were creating nanometer scale patterns on thin titanium film of 20nm thickness. Figure 2 demonstrates a digital photography of Alferov (a) and gray scale gradation in percentage values according to photo (b).

For height modulated oxide mask formation we used impulses of 1 millisecond with voltages between 6 to 10V according to the intensity of grey scale mask. There were used conductive cantilevers with W_2C coating [7].

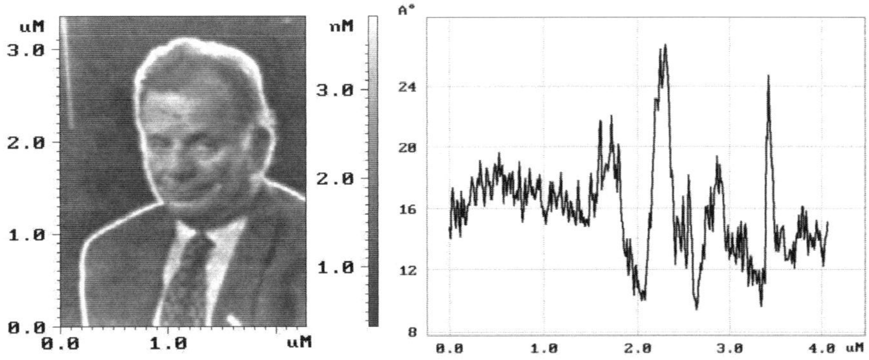

a) b)

Figure 3. AFM image of the titanium film surface after local anodic oxidation process a), and cross – section profile over the image b).

After scanning second time over same area which were treated by local anodic oxidation method the we observed nanometer scale oxide thickness formed on the titanium film according to loaded mask (see Figure 3 a). Oxide thickness variations are 0-3nm (see Figure 3 b).

Thus such example demonstrating the ability of whatever height modulated oxide creation by local anodic oxidation process based on any grey scale modulated mask and variation of the impulse duration or its value proportionally to gray scale intensity.

4. Conclusion

In paper the formation of a height-modulated dielectric film is proposed. It is obvious that if we need to create a dielectric mask with a different oxide thickness or a height-modulated film, it is necessary to vary the impulse duration or its value at different scanning field areas.

The correlation of the required pressing force on the conductive cantilever to the surface, which provides a close contact with the test sample, with material hardness and the thickness of the cantilever conductive coating, has been revealed.

The statements regarding height modulated dielectric mask formation on the solid state surfaces demonstrate the wide ability of local anodic oxidation by atomic force microscope for creation of different nanolelectronic devices.

Acknowledgments

This work was supported by the Russian Foundation for Basic Research (projects 16-38-00070).

References

[1] Dagata J A 1990 *J. Appl. Phys. Lett.* **56** 2001
[2] Belov A N, Gavrilov S A, Sagunova I V and Shevyakov V I 2010 *J. Semicond.* **44** 1709
[3] Lemeshko S, Gavrilov S, Shevyakov V et al 2001 *J. Nanotechnology* **12(3)** 273
[4] Lin J F, Tai C K and Lin S L 2006 *J. Appl. Phys. Lett.* **99** 1
[5] Lemeshko S V, Sagunova I V, Chapligin U A and Shevyakov V I 2015 *J. Proceedings of universities. Electronic* **20(3)** 188
[6] Lemeshko S, Bykov V, Saunin S, Roschin V and Gavrilov S 2003 *12th Int. Conf. on Scanning tunneling microscop* (Netherlands) 83
[7] Shevyakov V, Lemeshko S and Roschin V 1998 *J. Nanotechnology* **9** 352

IOP Conf. Series: Materials Science and Engineering **475** (2019) 012022 doi:10.1088/1757-899X/475/1/012022

InAlAs/InGaAs/InP heterostructures for microwave photodiodes grown by molecular beam epitaxy

D V Dmitriev[1], N A Valisheva[1], A M Gilinsky[1], I B Chistokhin[1], A I Toropov[1] and K S Zhuravlev[1]

[1]Rzhanov Institute of Semiconductor Physics, pr. ac. Lavrentieva, 13, 630090 Novosibirsk, Russia

Corresponding author's e-mail address: ddmitriev@isp.nsc.ru

Abstract. Pre-epitaxial (001)InP substrate cleaning, growth of $In_{0.52}Al_{0.48}As/In_{0.53}Ga_{0.47}As$ heterostructures lattice-matched to InP substrate by molecular beam epitaxy technique and manufacturing processes of microwave photodiodes with Schottky barrier are developed in this work.

1. Introduction

In microwave photonics systems, the powerful and high-speed photodetectors, capable to generate an output electrical signal with amplitude of about 1 V at frequencies of 10-100 GHz are required [13]. Different designs of powerful high frequency photodiodes for optical fiber communication and microwave photonics systems have been proposed in literature [4].

In our work the photodiode with Schottky barrier based on the InAlAs/InGaAs/InP heterostructure was selected for implementation because of the following advantages:

- Schottky barrier contact covers all photosensitive area that reduces the series resistance in comparison with that in the traditional photodiode with an annular contact with a hole in the middle for incoming radiation;
- Schottky barrier height of metal/InAlAs contact is high that reduces dark current of the photodiode;
- metal acts as a mirror for the radiation incoming through the substrate that considerably increases the absorption in the photodiodes;
- beam contact inductance can be adjusted to increase the maximum frequency of the photodiodes.

In present paper we focused on growth of InAlAs/InGaAs/InP heterostructure. To test heterostructures photodiodes have been manufactured and studied.

2. Experiment

The InAlAs/InGaAs heterostructures were grown on semi-insulating (001)InP substrates by molecular beam epiaxy (MBE) in Riber-32P and Compact-21T MBE multi-chamber systems equipped with solid-state molecular sources. Molecular flux of arsenic and metals (In, Ga, Al) was measured using an ionization gauge that was placed in front of sample during measurements. The substrate temperature was controlled by a single-wavelength infrared pyrometer. The growth process was *in-situ* monitored by reflection high-energy electron diffraction (RHEED) technique. The *ex-situ* evaluation of the heterostructures quality was performed by the atomic force microscopy (AFM) using a Solver P-47H (NT MDT) microscope with a vertical resolution about 0.5 Å, double-crystal X-ray diffraction and photoluminescence (PL) techniques. The values of concentration and mobility in layers of heterostructures were determined from the Hall effect measurement at magnetic field being B=0.2 T by the Van der Pauw method at a temperatures 77 K and 300 K. Ohmic contacts to the samples were obtained by the high-voltage discharge of indium foil imposed on a sample. For the fabrication of photodiodes, mesa isolation was performed by using BCl_3 plasma reactive ion etching and SiO_2 layer was deposited by plasma deposition to protect the surface and reduce leakage current. Au/Ge/Au/Ni metal stack was deposited by electron-beam evaporator for ohmic contact with thermal annealing at 385°C in hidrogen ambient. For forming Shottky barrier Ti/Au (20/200 nm) metal stack was deposited on the HCl etched InAlAs surface. From the measured current–voltage (I–V) curves, the Schottky barrier height, the ideality-factor, and the reverse saturation current were obtained.

3. Experimental results and discussion

3.1. Cleaning of InP substrates

Epi-ready InP substrates were cleaned by annealing in a high vacuum chamber in order to remove the oxide layer and form atomically clean surface. First, the substrate was annealed at a temperature of 200-250°C below the temperature of noncongruent evaporation of InP being 325°C in a special MBE chamber separated by a vacuum gate of the growth chamber. Volatiles and water were removed from the substrate surface without contamination of the growth chamber. The subsequent high-temperature annealing happened in growth chamber. Then oxide layer was removed at a temperature above the congruent evaporation temperature under the arsenic flux with beam equivalent pressure (BEP) $F_{As}=5\cdot 10^{-5}$ Torr in order to prevent formation of In droplets. At a temperature of 350°C (2×3) surface reconstruction was formed. At a temperature of 450°C (2×6) reconstruction appeared. To the best of our knowledge, this reconstruction is reported for the first time. We assume that it is caused by the replacement of the phosphorus atoms of arsenic atoms. With further increase in temperature (2×6) reconstruction faded, while (4×2) reconstruction appeared at 540°C (see Figure 1) indicating the evaporation of elements of the group V (P and As). Note that (2×8) surface reconstruction was observed earlier under a phosphorus flux in the same temperature range [5]. The emergence of (4×2) reconstruction, increasing the intensity of the fractional stripes, decrease in the background intensity indicates the evaporation of the oxide layer [6]. During cooling of clean surface (2×4) reconstruction was observed. Annealing was accompanied by the desorption of phosphorus atoms which partially replaced by arsenic atoms [7–9]. This process led to the formation of InAs islands with height and lateral size of about 0.84–1.16 nm (2-3 monolayers) and 100 nm, correspondingly. Heterostructures were grown on substrates annealed at 540°C.

3.2. Growth of InAlAs u InGaAs layers

To establish growth conditions of $In_xAl_{1-x}As$ (x_{In} close to 0.52) and $In_yGa_{1-y}As$ (y_{In} close to 0.53) layers lattice-matched with InP the substrate temperature and arsenic flux (BEP) varied between 460-580°C and $(1\text{-}10)\cdot 10^{-5}$ Torr, accordingly. Two methods were used to control the molecular flux ratio of metals and arsenic coming to the surface growth.

- Measurement of BEP by an ionization vacuum gauge considering the ionization coefficient of each chemical element. Indications of the gauge at the opened and closed source shutter were measured. Calculation from these values is a BEP each material.
- Measuring the rate of growth, using the time oscillation in the intensity of RHEED strip. The growth rate of ternary alloy is determined by the total flux of metals, the flux of each metal was evaluated in growth process of binary compounds. These data were used to define the composition and thickness of the alloy layers.

According to the X-ray diffraction (XRD) data, the best structural parameters of InAlAs layers were obtained at 520-535°C for BEP_{As} of $(1.5\text{-}2)\cdot 10^{-5}$ Torr. Measured and calculated positions of (004) XRD peaks of the InAlAs layer good match. The full width at half maximum (FWHM) of double-crystal X-ray rocking curve was as low as 20" , that is very close to the value of 13" for InP substrate, as illustrated in Figure 2. The Figure 3 shows PL spectra at 77 K and 300 K from the InAlAs

IOP Conf. Series: Materials Science and Engineering **475** (2019) 012022 doi:10.1088/1757-899X/475/1/012022

layer grown at 535°C. The 77K PL spectrum comprises a single band peaked at 1.508 eV with a FWHM of 27 meV that originated from band-to-band transitions. In contrast to PL from the AlInAs layers grown at lower temperatures no impurity-related bands have been observed.

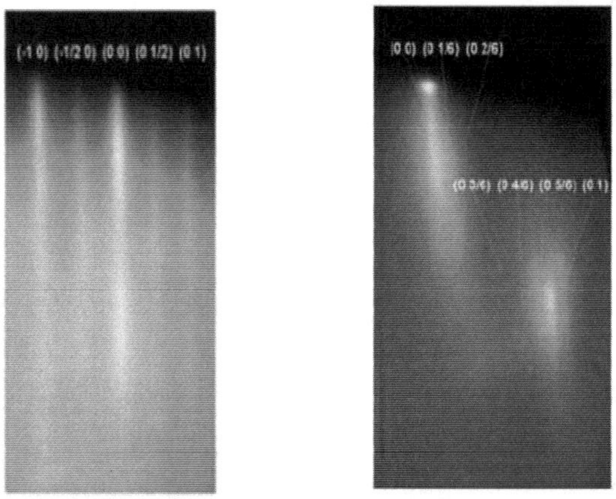

Figure 1. RHEED images of InP substrates annealing under the arsenic flux with BEP_{As} of $5 \cdot 10^{-5}$ Torr, (2×6) reconstruction at a temperature of 450°C. Image showing the 2-by and 6-by periodicity in the [110] and [$\bar{1}$10] azimuth direction, respectively.

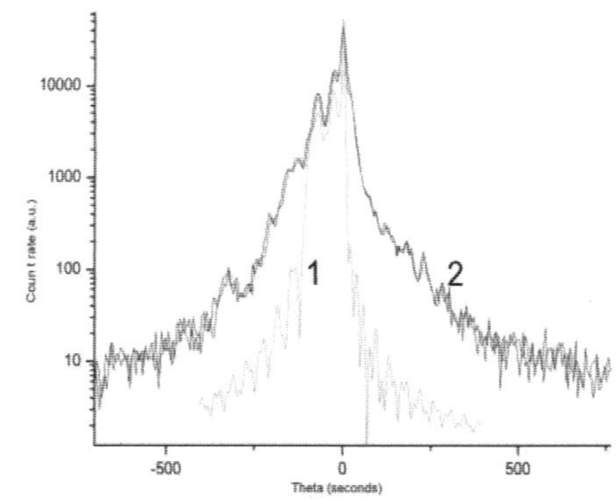

Figure 2. Calculated (1) and experimental (2) X-ray rocking curve (Cu Ka, 004 reflex) of the $In_{0.52}Al_{0.48}As$ layer grown at a temperature 535°C and BEP_{As} $1.5 \cdot 10^{-5}$ Torr.

Figure 4 demonstrates the AFM image of the InAlAs layers grown at 535°C and BEP_{As} $1.5 \cdot 10^{-5}$ Torr with monoatomic steps, evidencing in a two-dimensional growth mode. The root mean square (RMS) value of the surface roughness in the 1x1 μm^2 area is about 0.113 nanometers.

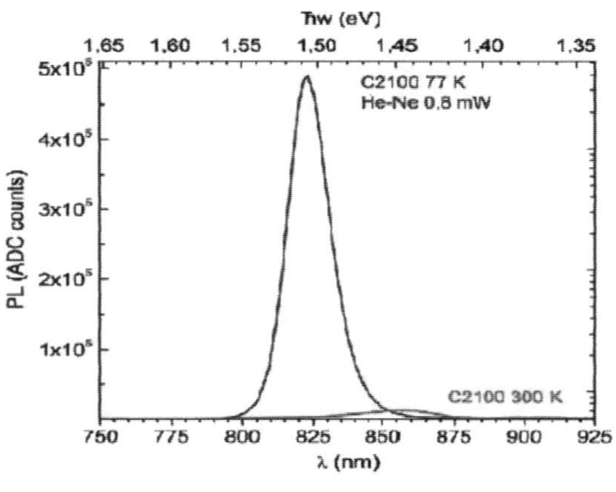

Figure 3. 77 and 300 K PL spectra.

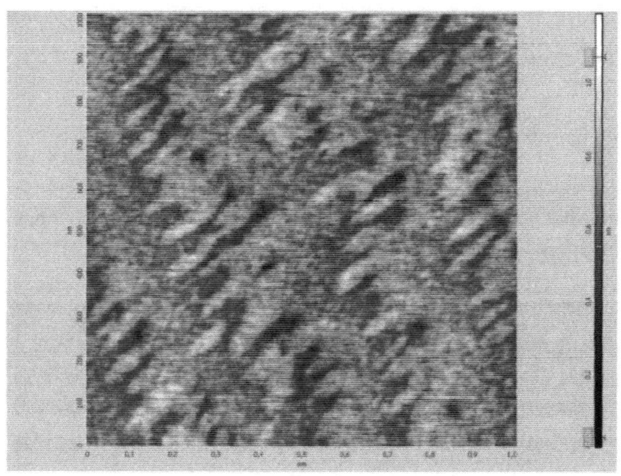

Figure 4. 1x1 μm^2 AFM image of $In_{0.52}Al_{0.48}As$ $In_{0.52}Al_{0.48}As$ layer grown at a temperature layers grown at a temperature 535°C and BEP_{As} 535°C and BEP_{As} $1.5 \cdot 10^{-5}$ Torr. $1.5 \cdot 10^{-5}$ Torr.

Dependence of carrier concentration and mobility of $In_{0.52}Al_{0.48}As$ layers on growth conditions has been investigated. It was found that the electron concentration decreases when the growth temperature reduces.

The following conditions were chosen for growth of $In_{0.53}Ga_{0.47}As$ layers: a 480°C growth temperature and arsenic BEP $3 \cdot 10^{-5}$ Torr. In these conditions two-dimensional growth occurs, the RMS value is 0.5–0.7 nm, i.e. 2-3 monolayers. XRD evidences in structural uniformity of these layers.

The electron concentration and mobility at 300 K was about $1.45 \cdot 10^{15}$ cm^{-3} and 6600 cm^2/(V·sec), respectively. At 77 K epitaxial $In_{0.53}Ga_{0.47}As$ layers become non-conducting.

3.3. Growth of InAlAs/InGaAs/InP heterostructure

The layer consequence for microwave photodiode heterostructure is presented in table 1.

Table 1. InAlAs/InGaAs/InP heterostructure for microwave photodiodes with Schottky barrier.

$In_{0.53}Ga_{0.47}As$	30 nm	Protect layer
$In_{0.52}Al_{0.48}As$	30 nm	Barrier layer
InGaAlAs	50 nm	Gradient layer
$In_{0.53}Ga_{0.47}As$	600-1500 nm	Undoped absorbing layer, thickness depends on operating frequency
$In_{0.53}Ga_{0.47}As$:Si $5 \cdot 10^{18}$ cm^{-3}	50 nm	Absorbing n$^+$ layer
$In_{0.52}Al_{0.48}As$:Si $5 \cdot 10^{18}$ cm^{-3}	300 nm	Contact layer
InP (001)	400 μm	Substrate

At the growth of the InAlAs/InGaAs/InP heterostructure each layer was grown at different temperature, namely, InGaAs layers were grown at 480°C whereas InAlAs layer was grown at a temperature above 500°C. To conduct continuous process that excludes capture of undesirable background impurity in chemically active Al-containing layers and does not overheat InGaAs layers, growth of InAlAs layer was divided into several stages. Growth started at 480°C, and then the temperature was gradually raised to about 530°C, finally the temperature was lowered to 480°C. This approach allows to grow heterostructure in a two-dimensional growth mode with RMS lesser 0.3 nm.

Microwave photodiodes with a various diameters of 200, 150, 100, 75 and 50 μm of Shottky contact have been fabricated. Devices showed typical diode *I-V* characteristics. For 50 m photodiodes the Schottky barrier height 0.787 eV, the ideality-factor 1.07 and the reverse current less $2 \cdot 10^{-9}$ A was obtained.

4. Conclusion

The technology of pre-epitaxial vacuum cleaning of (001) InP substrates and MBE growth of multilayer heterostructures for microwave photodiode were considered. Optimum conditions for growth of InAlAs/InGaAs heterostructures were established. The suitability of heterostructure for manufacturing of photodiodes was demonstrated.

This work was supported by the RFBR grants nos.14-29-08124.

5. References

[1] Karve G, Zheng X, Zhang X, Li X et al 2003 *IEEE J. Quantum Electronics* **39** 1281
[2] Mun S C L T, Tan C H, Dimler S J et al 2009 *IEEE J. Quantum Electronics* **45** 566
[3] Tan L J J, Ong D S G, Ng J S et al 2010 *IEEE J. Quantum Electronics* **46** 1153
[4] Chistokhin I B and Zhuravlev K S 2015 *Advances in Applied Physics (in Russian)* **3(1)** 85
[5] LaBella V P, Ding Z, Bullock D W et al 2000 *J. Vac.Sci. Technol.* A **18** 1492
[6] Davies G, Heckingbottom R, Ohno H et al 1980 *Appl. Phys. Lett.* **37** 290
[7] Li C H, Li L, Law D C, Visbeck S B and Hicks R F 2002 *Phys. Rev.* B **65** 205322
[8] Schmidt W G and Bechstedt F 1998 *Surf. Sci.* **409** 474
[9] Krzyzewski T J and Jones T S 2008 *Phys. Rev.* B **78** 155307

Effect of thin heterogeneous functional nanolayers on electron transport in InGaAs-based quantum wells with high electron density (a review)

I S Vasil'evskii[1]

[1]National Research Nuclear University «MEPhI», Russia, 115409, Moscow, Kashirskoe sh. 31

Corresponding author's e-mail address: ivasilevskii@mail.ru

Abstract. A prospective improvement of InGaAs-based quantum well heterostructures design is possible towards the quantum treatment of electron subbands states and electron transport processes in the real structures with thin spatially inhomogeneous functional nanolayers. Insertion of nanosized GaAs, AlAs and InAs layers into quantum well/spacer/barrier layers or applying a graded $In_yGa_{1-y}As$ channel can enhance electron transport properties and gives a novel degree of freedom for the quantum structures engineering. The review and original study for PHEMT/GaAs, HFET/GaAs and InP HEMT structures are discussed.

1. Introduction

Compound semiconductor heterostructures become widely last thirty years and provided a unique functionality and performance due to the quantum engineering. Among the most important application the field effect transistors based on high electron mobility transistor (HEMT) structures became a fundamental for microwave nanoelectronics [1]. Unless InGaAs quantum well HEMT structures still dominate in the compound semiconductors for microwave monolithic integrated circuits (MMIC), the insight in electron transport processes in such structures aren't so deep as for well elaborated "classical" AlGaAs/GaAs HEMT. Researchers sought rather to benefit from these structures for the microwave applications than to go deeper in fundamental electron processes. In particularly it is reflected in empirical approach for the set of physical relations, for example velocity-field, bandgap-temperature and others. Since initial stage of detailed study of AlGaAs/GaAs HEMT left away and condensed matter theoreticians attention moved to the spin related phenomena and strong correlated systems, leaving a lot of questions for the detailed insight. The practical heterostructure design turned to deeper InGaAs- based quantum well structures with different InAs content that usually have a higher concentration of two-dimensional electron gas. Electron transport peculiarities in such structures haven't been investigated in details. InGaAs based quantum wells have a different basic types dependent on InAs composition and substrate type: AlGaAs/InGaAs pseudomorphic HEMT (PHEMT, ~20% InAs), lattice matched to InP InAlAs/InGaAs HEMT (53% InAs) and pseudomorphic QW HEMT on InP (60-80% InAs) [2]. The structures also differ by dopant arrangement (one side or double side delta-doping) and nowadays steady trend of double-side doping strategy appears. Power performance along with cutoff frequency increase in HEMT application require the further development of widely used InGaAs channel HEMTs [3] while GaN based structures are still more expensive and technologically tough.

Another apparent trend towards the increase of MMIC operation frequency is the increase of InAs composition in QW channel. Heterostructure design follows the three dimensional scaling, because one should bring the channel closer to the surface to maintain electrostatic control together with the decrease of gate length.

Due to the surface potential and high charge density in modern HEMT structures band profile is affected by strong electric field. In turn it causes the changes both electron confinement and subband structure at low electric field and real space transfer of hot carriers [3, 4]. Complicated quantum effects as state hybridization can take place [5].

Additional degree of freedom for quantum well HEMT structures engineering gives a composite functional nanolayers such as composite quantum well, spacer and barrier layers. The major importance for such structures must be paid for multiple subband electron transport, effective mass engineering and self-consistent electron wavefunction spatial profiles.

The present paper gives the state-of-the art overview of different InAs content structures with functional nanolayer insertions and spatially inhomogeneous layers in PHEMT and HFET $Al_xGa_{1-x}As/In_yGa_{1-y}As/Al_zGa_{1-z}As$ on GaAs, HEMT and PHEMT $In_{0.52}Al_{0.48}As/In_yGa_{1-y}As/In_{0.52}Al_{0.48}As$ on InP. The considered approaches employ either binary AlAs and GaAs inserts or ternary $In_yGa_{1-y}As$ and $Al_zGa_{1-z}As$ layers with compositional gradient.

2. Composite spacer with AlAs nanoinserts for high density 2DEG in HEMT and PHEMT quantum well structures

High doping through a spacer usually applied to obtain high density of two dimensional electron gas in HEMT quantum wells. Resulting charge separation produces dipole electrical field that lowers the conduction band energy around the doped wideband barrier and also bends a quantum well profile. This negative effect on the one hand reduces quantum well effective depth and for the other, gives rise to tunnel penetration of electron wavefunction from QW to the barrier. Interaction of electron states is manifested in subband energy level anticrossing with increasing spacer thickness accompanied by electron mobility crossing [6]. As this effect has tunnel origin, an insertion of 2 nm AlAs barrier in AlGaAs spacer can effectively destroy the tunnel states' coupling and separates electron mobility to higher one for QW-confined state and lower one associated with V-shaped potential well of ionized donors. For the shallow $Al_{0.11}Ga_{0.89}As/GaAs/Al_{0.11}Ga_{0.89}As$ QW 3.4 fold electron mobility increase have been obtained at low temperature (T=77 K) [6]. Note that single thin AlAs doesn't change an overall electron spatial distribution and weakly affects the subband energies.

Applying double AlAs barrier insertions surrounding the AlGaAs:δ-Si donor layer causes substantial energy shift for the electron states associated with V-potential well in the QW barrier and completely crowds out electron density from the barrier. Temperature dependence of electron sheet concentration n_s showed a decreased temperature variation due to an increase of donor ionization energy level by such insertions. For 1δ-PHEMT $Al_{0.15}Ga_{0.85}As/In_{0.15}Ga_{0.85}As/GaAs$ quantum well δ-Si doped GaAs donor layer was applied, as shallow Si ionization energy provides very high electron concentration. However, introduction of double AlAs nanolayers around δ-Si:GaAs sufficiently decreases temperature sensitivity of Hall electron concentration from 37% to 12% and enhances electron mobility in the sample. Low temperature magnetotransport data for the two samples 261 and 262 with simple spacer and with composite spacer containing double 2 nm AlAs insertions around δ-Si layer depicted in Figure 1 a) and b), correspondingly. It can be seen that for composite spacer structure 262 magnetoresistance demonstrates approximately constant monotonic component and Shubnikov-de Haas oscillation started at a smaller magnetic field. For the sample 261 magnetoresistance monotonic component is positive and quadratic that is typical for multiple subband occupation. This indicates that in composite spacer sample electron states considered in [7] are moved away from delta-doped layers and electron mobility in ground subband in QW is increased. Fourier analysis of SdH oscillation determined electron density in QW ground subband. Analysis showed a slight increase of electron density in QW for sample 262 and the absence of a parallel conduction, whereas for sample 261 electron density in QW accounted as only 65% at high temperatures. Sample

262 has n_s=1.55·10^{12} cm^{-2} and μ_H = 6720 cm^2/(V s) and μ_H = 25 770 at T=300 K and T=4.2 K, respectively.

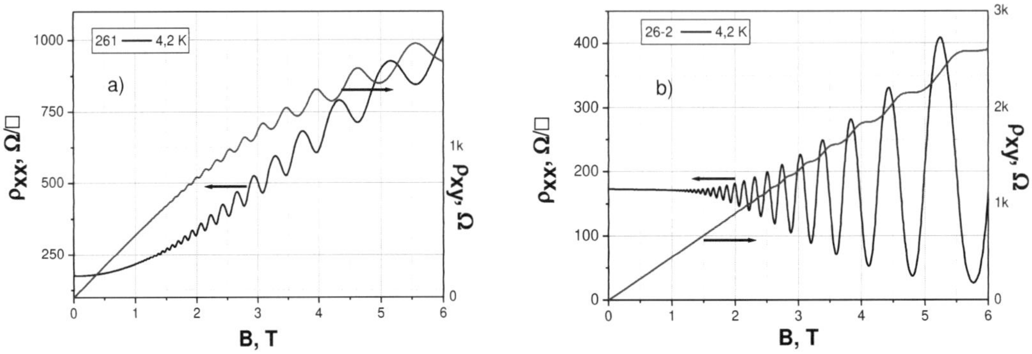

Figure 1. Magnetoresistance and Hall effect in high magnetic field at T=4.2 K for PHEMT QW with simple spacers (a) and composite spacers containing double AlAs insertions (b).

Thin directly δ-Si doped AlAs donor layer in PHEMT $Al_{0.24}Ga_{0.76}As/In_{0.2}Ga_{0.8}As/GaAs$ QW showed mobility enhancement up to μ_H = 30 000 cm^2/(V s) and 33 000 cm^2/(V s) at T=77 K and T=4.2 K, respectively [8], that are comparable values for InP HEMT structures. On the other hand, doping efficiency are shown to be 3-4 times smaller in comparison to GaAs so donor concentration must be sufficiently increased in AlAs donor layers to provide n_s~1.5÷2·10^{12} cm^{-2}.

AlAs thin barriers grown in QW spacers allow also to avoid a real space transfer of hot electrons the towards wideband barriers in high electric field, that is rather important for microwave transistors. Another approach presented in papers [3, 9] for DA-PHEMTs utilize an engineering of built-in electric field by combination of donor and acceptor compact doping. In this case tunneling from QW to wideband AlGaAs barrier are suppressed by high electrostatic potential in acceptor layers. Substantial increase of transistor gain has been reported in DA-PHEMTs compared with ordinary PHEMT structure [3].

3. Doped GaAs nanolayers for extremely high density 2DEG in HFET quantum well structures

For power applications heterostructure field effect transistor structures (HFET) are used with directly doped channel. This eliminate any parallel conduction and allow one to increase AlAs content in the Schottky barrier as high as 30-35% to depress a leakage current and provide large modulation of gate signal [10]. On the contrary this structure layout breaks the principle of modulation doping and towards the electron concentration in the range n_s~3÷5·10^{12} cm^{-2} HFET structures demonstrate very low electron mobility μ_H ~ 1 500 cm^2/(V s). A compromise approach was presented in our previous work [11] by the compound channel structure $Al_{0.38}Ga_{0.62}As/In_{0.15}Ga_{0.85}As/Al_{0.38}Ga_{0.62}As$ with additional 9 nm GaAs sublayers that served as sub-barriers at the QW heterointerfaces. Redistribution of Si doping from the uniform doped QW to delta-doping in two GaAs nanobarriers have been studied and an enhancement of QW conductivity and electron mobility have been found. Effect is explained by suppression of electron amplitudes at the sub-barrier area and decrease of ionized impurity direct electron scattering when the dopants are partially placed aside of channel center. A combination of n_s = 1.37·10^{13} cm^{-2} and μ_H = 1520 cm^2/(V s) have been obtained in such structure.

4. Compositionally graded PHEMT quantum wells

For the double-side δ-Si doped 2δ-PHEMTs $Al_xGa_{1-x}As/In_yGa_{1-y}As/Al_xGa_{1-x}As$ an increase of doping concentration allows to obtain n_s~3÷5·10^{12} cm^{-2} but accompanied by a decreased electron mobility. One of the reasons is the decrease of effective QW depth due to rise of parabolic electrostatic potential barrier at the QW center. This barrier deforms electron wavefunction (WF) by moving electron density towards the heterointerfaces. The other reason for decreased μ is shrinking energy difference between the ground and first excited subband. As a result, electron density in the ground subband saturated with increasing dopant concentration. This unwanted effect can be compensated by deliberate growth of graded composition layer $In_yGa_{1-y}As$ taking into account $E_g(y)$ dependence. This approach has been realized by MBE with relatively slow growth rate and very high compositional gradient up to 2%/nm to compensate linear (for 1δ-PHEMTs) and parabolic (for 2δ-PHEMTs) band bending at QW [12]. For the one-side doped PHEMTs such a grading of quantum well bottom doesn't vary electron subband energies but moves the centroid of 2DEG towards the QW center. For 2δ-PHEMTs effective QW depth (which can be defined as energy separation between ground electronic subband and the minimum for barrier potential) increased in the QW with elevated InAs content in the center, even though mean InAs content in $In_yGa_{1-y}As$ layer remained. The structure with optimal InAs profile that produced most rectangular profile shown to have both a highest mobility and conductivity (latter is increased by 25% compared with the standard structure).

5. Composite quantum wells with InAs single and multiple inserts

Increase of InAs composition in $In_{0.52}Al_{0.48}As/In_yGa_{1-y}As/In_{0.52}Al_{0.48}As$ HEMT structures grown on InP substrates realized by the two approaches. First is to growth pseudomorphically strained QW with uniform InAs composition ($0.53< y <0.8$), however, even for y=0.8 substantial thickness limitation due to strain restricts QW thickness to less than 10 nm. The second technique is compound quantum well (CQW) with additional InAs sub-channel. These structures have demonstrated great performance for short channel microwave transistors and circuits [13, 14]. Nevertheless, such structures aren't deliberately explored from the quantum states and electron transport peculiarities, concerning e.g. negative effects of CQW thickness shrinkage, comparison of CQW with the different numbers of InAs inserts. For single InAs insert it is known that WF of ground state grows at the insert place and leads to CQW effective thickness shrinkage. Accordingly to the tunnel effect we have supposed that multiple InAs inserts could conserve relatively wide composite QW. Self-consistent calculation for single QW and composite CQW, containing one, two and three InAs inserts has been performed. Conduction band and electron wavefunctions profiles are presented in Fig 2 for single QW, two and three 2 nm InAs inserts CQW. It can be seen that double InAs inserts slightly modulate the profile of ground state wavefunction, whereas tree inserts CQW demonstrates the shift of 2DEG centroid towards the upper heterointerface. Thus right CQW with 3 inserts edge substantially shifts upward and WF amplitude at the third insertion becomes rather small. This indicates that 3-insertion CQW isn't effective, but 2-insertion CQW is quite attractive. Energy levels in the CQW with different InAs total thickness have been calculated and shown on Fig. 3. Three models are presented - single InAlAs/InAs/InAlAs QW, and InAlAs/InGaAs/InAlAs - based CQW with one and two InAs inserts.

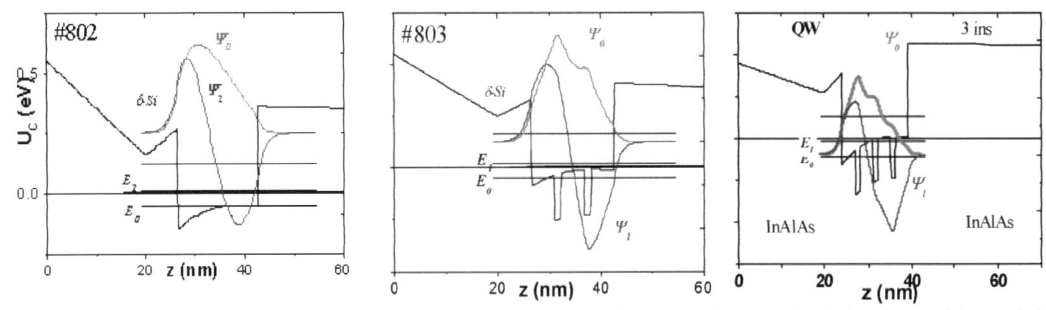

Figure 2. IR-transmission spectra of experimental samples (a). The result of decomposition of the absorption spectrum for the sample deposited at 800°C (b).

To account for the band structure effects on electron transport properties, electron mobility calculation have been performed for interface roughness scattering mechanism. The mobility dependence on InAs total thickness is shown in Fig. 4. It can be seen that for the interval L_{ins}=1÷4 nm the highest μ corresponds to the double insertion CQW, and single deep InAs QW demonstrates a lower μ. The main reason is that the different quantum structures have the different energy level variation due to insertion thickness deviation. The determined dependence of $\partial E/\partial L_{ins}$ on L_{ins} has the form $\partial E/\partial L_{ins} \sim L_{ins}^{-k}$ where $k\sim$ 2.4 for SQW, $k\sim$ 1.0 for the one-insertion CQW and $k\sim$ 0.6 for the double insertion CQW. Note that for infinitely deep QW the exponent should be $k\sim$ 3. This implies that for double insertion CQW quantum well has a large effective thickness and less sensitive to individual insert roughness. Also for the double insertion CQW energy difference between the two lower subbands remains nearly constant in the practical range of 1 nm<L_{ins}<5 nm, indicating that electron states are not localized at individual inserts in contrast to the case of a single insertion CQW.

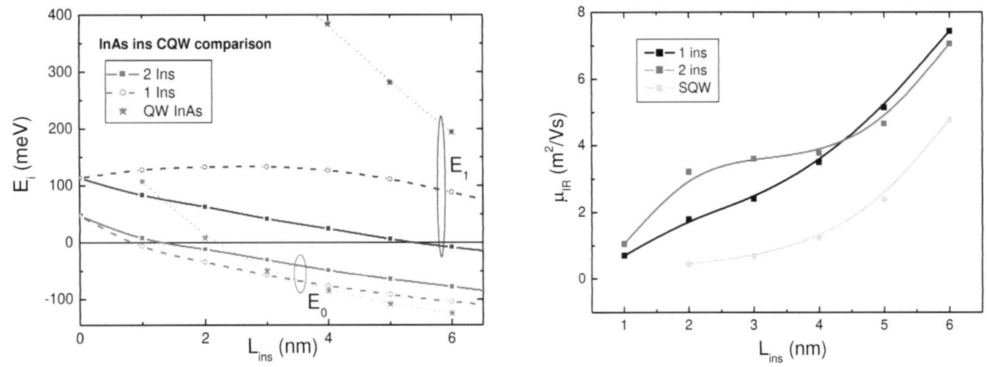

Figure 3. IR-transmission spectra of experimental samples (a). The result of decomposition of the absorption spectrum for the sample deposited at 800°C (b).

Experimental for the two InAs insertion CQW have shown indeed some decrease of electron low field mobility [15], probably because lateral roughness and mechanical deformation distribution played an additional important role in CWQ crystalline perfection and scattering processes. However in high electric field clear dependence of high field mobility and electron velocity and also saturation field value have been reported depending on either one or two InAs insertion presence and GaAs sub-barriers at the CQW edges. The main reasons of such effects are effective mass tuning and also optical phonon modes selection by an insertions [16, 17], as it was predicted by numerous works devoted to optical phonon confinement and scattering form-factor adjustment by thin heterogeneous layers [18, 19]. Thus InAs and GaAs nanosized insertions rather than AlAs wideband inserts served as electron-phonon scattering tuning barriers, because AlAs too much outrages electron subband structure being introduced in QW area.

6. Conclusion

A number of new approaches that using spatially inhomogeneous functional nanolayers in InGaAs channel-based HEMT quantum wells are discussed in the context of the influence on electron states, quantum treatment of electron scattering processes. Insertion of nanosized GaAs, AlAs and InAs layers into basic functional layers as quantum well/spacer/barrier or applying a graded $In_yGa_{1-y}As$ channel can enhance electron transport properties and gives a novel degree of freedom for the quantum structures engineering.

This work was supported by MEPhI Global competitiveness program and MEPhI Shared-Use Equipment Center «Heterostructure microwave electronics and wide bandgap semiconductor physics».

References

[1] Alamo J A 2011 *Nature* **479** 317
[2] Marsh P F, Chu S L, G Lardizabal S M et al 1999 *IEEE MTT-S International Microwave*

Symposium Digest **4**
[3] Lukashin V M, Pashkovskii A B, Zhuravlev K S et al 2014 *Semicond.* **48** 666
[4] Zou J, Abid Z, Dong H and Gopinath A 1991 *Appl. Phys. Lett.* **58** 2411
[5] Vasilevskii I S, Kulbachinskiĭ V A, Lunin R A et al 2007 *J. Exp. Theor. Phys.* **105** 174
[6] Vinichenko A N, Gladkov V P, Kargin N I et al *Semicond.* **48** 1619
[7] Kulbachinskii V A, Vasil'evskii I S, Lunin R A et al *Semicond. Sci. Technol.* **22** 222
[8] Vinichenko A N and Vasil'evskii I S 2016 *IOP Conf. Ser.: Mater. Sci. Eng.* **151** 012037
[9] Gulyaev D V, Zhuravlev K S, Bakarov A K et al 2016 *J. Phys. D: Appl. Phys.* **49** 095108
[10] Lin Y C, Chang E Y, Yamaguchi H et al 2007 *IEEE Trans. Electron. Dev.* **54** 1617
[11] Khabibullin R A, Vasil'evskii I S, Galiev G B et al 2011 *Semicond.* **45** 1321
[12] Vasil'evskii I S, Vinichenko A N, Grekhov M M et al 2014 *Semicond.* **48** 1226
[13] Nakayama T, Miyamoto H, Oishi E and Samoto N 1996 *J. Electr. Mater.* **25** 555
[14] Mei X, Yoshida W, Lange M et al 2015 *IEEE Electron Device Lett.* **36** 327
[15] Kulbachinskii V A, Yuzeeva N A, Galiev G B et al 2012 *Semicond. Sci. Technol.* **27** 035021
[16] Požela J, Požela K, Jucienė V et al 2011 *Lithuanian J. Phys.* **51** 270
[17] Šilenas A, Požela J, Požela K et al 2013 *Semicond.* **47** 372
[18] Požela J, Požela K, Raguotis R and Jucienė V 2011 *Semicond.* **45** 761
[19] Požela K, Šil˙enas A, Požela J et al 2012 *Appl. Phys. A* **109** 233

IOP Conf. Series: Materials Science and Engineering **475** (2019) 012025 doi:10.1088/1757-899X/475/1/012025

Formation and study of PbS quantum dot films with different ligands

A E Aleksandrov[1], M A Zvaigzne[1], I L Martynov[1], D A Lypenko[2], A R Tameev[2] and A A Chistyakov[1]

[1] National Research Nuclear University MEPhI (Moscow Engineering Physics Institute), Kashirskoe shosse 31, 115409 Moscow, Russian Federation
[2]A.N. Frumkin Institute of Physical Chemistry and Electrochemistry of the Russian Academy of Sciences, 31-bld.4 Leninsky Prospect, Moscow 119071, Russia

Corresponding author's e-mail address: Alexeyalex93@mail.com

Abstract. We study the influence of various organic molecules of ligands on the optical properties of the condensates of lead sulfide quantum dots (QD) and photoconductivity of diode structures made on their basis. It is shown that the decrease of the molecule length of surface ligands by a factor of 4 leads to the significant increase in the conductivity, and the emergence of photocurrent. We also demonstrate that the thickness of the photoactive QD layer plays an important role in the performances of photovoltaic cells.

1. Introduction

Due to the dependence of the absorption and luminescence spectra on the size of quantum dots (QDs), their high photostability and wide absorption spectrum, QDs are considered to be a promising material for use in optoelectronic devices. QDs based on lead chalcogenide, which are capable of absorbing light in the infrared spectrum region, are of great interest [1-3]. Photovoltaic cells, up-conversion devices and IR photodetectors can be created on their basis. However, without achieving a low defect concentration and good charge carrier transport in QD layers, it is difficult to obtain high-efficiency values of such devices. The important factor determining the effectiveness of QDs application is a presence of the layer of organic molecules (ligands) on the nanocrystal surface. The selected molecules should not form high barriers for the charge carriers tunneling, but at the same time, they should spatially separate the nanocrystals in order to avoid their aggregation. Since charge transfer between quantum dots occurs via "hopping" mechanism [4], the important factor affecting the electrophysical properties of thin QD films is the ligand molecule length which determines the distance between QDs. Ligands should also provide passivation of uncompensated bonds on the nanocrystal surface since they can form traps for charge carriers. At the same time, it is promising to use ligands, which are able to chemically bind QDs together, forming a uniform covalent nanostructure.

In this paper, we studied the effect of various organic ligands on the electrophysical properties of QD layers.

2. Sample fabrication

The used PbS quantum dots were synthesized by the colloid chemistry method. To produce a QD film, 50 μl of QD solution in toluene with concentration of 10 mg /ml was applied by automatic pipette onto a pre-cleaned glass substrate and then the substrate was rotating for 30 seconds at an angular rate of 1500 rpm with spin-coating method. After that 50 µl of ligand molecules solution (1 mol/l of hydrazine in methanol; 0.1 mol /l of HDT (1,6-hexanedithiol) in acetonitrile) was applied onto the film and after 30-second exposure the film was rotating for 30 seconds at angular rate of 1500 rpm. To remove the excess of ligand molecules from the film surface, a suitable for each ligand solvent was applied on it, and then the film was rotating for 30 seconds at angular rate of 1500 rpm. The described procedure was repeated 5 times [5, 6].

Initially, as a result of the synthesis, the QDs surface has been covered with a layer of oleic acid molecules (OA), which length is approximately 1.5 nm. After ligand replacement, the QDs surface was covered with a layer of hydrazine and 1.6-hexanedithiol (HDT), the molecules length of which are approximately 0.5 nm and 1 nm, respectively. Such choice of ligands is caused by small length of their molecules and the high affinity of thiol groups to sulfur molecules.

Also, on the base of QD films photovoltaic elements were produced (ITO/PEDOT: PSS/PbS/ZnO/Al). Polymer PEDOT: PSS was deposited by the spin-coating method on a transparent glass substrate, covered with an ITO layer. The substrate angular rate was 2000 rpm and the layer thickness was 30 nm. After that, the sample was dried at the temperature of 60°C for several hours. Then a layer of QDs was deposited, and the ligands were replaced according to the procedure described above. The number of QDs layers in the structures was varied. The subsequent buffer layer of ZnO QDs was deposited by the spin-coating method at an angular rate of 2000 rpm, the layer thickness was 30 nm. Due to the quantum-size effect, the ZnO QDs were chosen so that the energy levels of the buffer layer do not create a barrier and do not prevent the diffusion of charge carriers to the contacts. After applying all the layers aluminum contacts were sprayed on each structure through the mask.

3. Results and discussion

Figure 1 shows the IR spectra of films with different ligands obtained with the IR spectrometer SHIMADZU IRaffinity-1S. The vibrational peaks characteristic only for a specific molecule are indicated for each ligand. In this case, the distinguishable peak for oleic acid ($C_{18}H_{34}O_2$) is one at 3080 $см^{-1}$, corresponding to stretching vibrations of the bond =C-H; for 1.6-hexanedithiol ($C_6H_{14}S_2$) molecule – the peaks corresponding to stretching vibrations of the bond S-H (2620 $см^{-1}$) and C-S (670 $см^{-1}$); for hydrazine (N_2H_4) – the peak of stretching vibrations (3280 $см^{-1}$) and deformation vibrations (1590 $см^{-1}$) of the N-H bond. Thus, the presence of the above peaks confirms the replacement of the original ligands with the new ones.

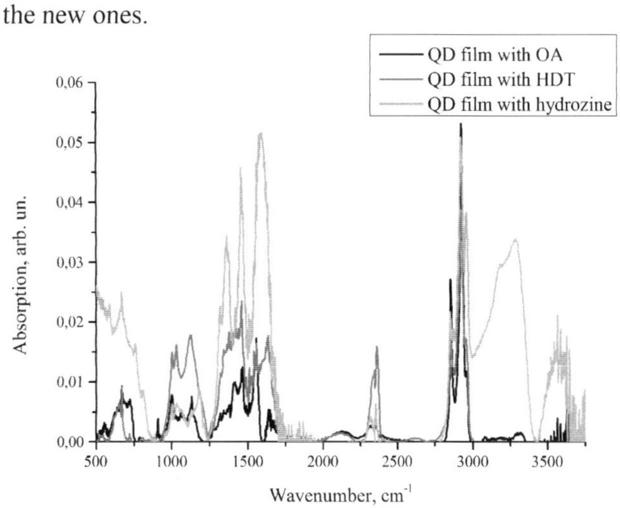

Figure 1. IR spectrum of QD films with oleic acid (OA), 1.6-hexanedithiol (HDT) and hydrazine as surface ligands.

For all film samples, conductivity was measured in the dark and under irradiation by a wide-range halogen lamp Ocean Optics LS-1. For this purpose, thin aluminum contacts were sprayed onto the surface of the films through a mask. On the basis of the I-V characteristics measurements, the relative conductivity has been calculated for each sample. Figure 2 shows that the conductivity increases exponentially with decreasing of ligand molecule length. This result is in good agreement with the theory of the hopping mechanism of charge transfer between QDs in a film, within which the mobility contains the exponential factor **exp(-2γl)**, where γ- is the inverse radius of wave functions localization, **l**- is the tunnel jump length (ligand molecule length).

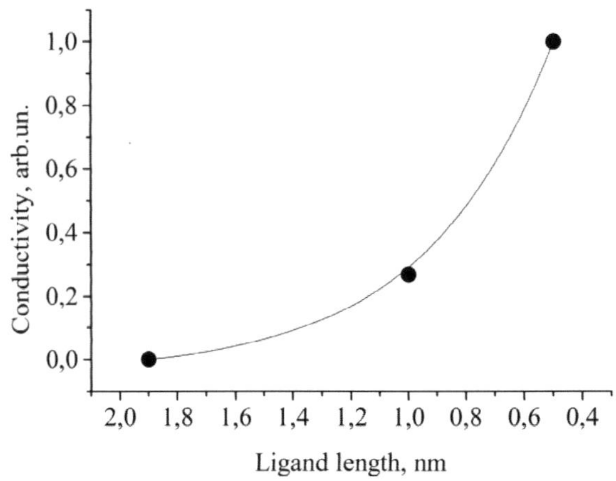

Figure 2. The relative change of QD film conduction with the decrease of ligand length.

Thus, the replacement of relatively long molecules of the initial OA ligands by shorter ones that are capable of chemically bonding QDs to each other leads to an efficient charge transfer between the QDs. In this case, conductivity is largely determined by the ligand molecule length, rather than its chemical nature.

Using the proven technique of manufacturing QD films, photovoltaic elements on the basis of QDs with substituted ligands layers (ITO/PEDOT:PSS/PbS/ZnO/Al) were made and the dependence of the I-V characteristics on the nanocrystals layer thickness was studied. The thickness was varied by changing the number of QD layers. Structures with 3, 5, 6 and 10 QD layers were made. All the samples demonstrated pronounced diode characteristics (figure 3). The measured representative I-V characteristics of the samples under irradiation are shown at figure 4. Samples with six QD layers showed higher values of efficiency compared to the others. Thereby, the optimal thickness of the photovoltaic structure is about 70-90 nm (6 layers of QDs).

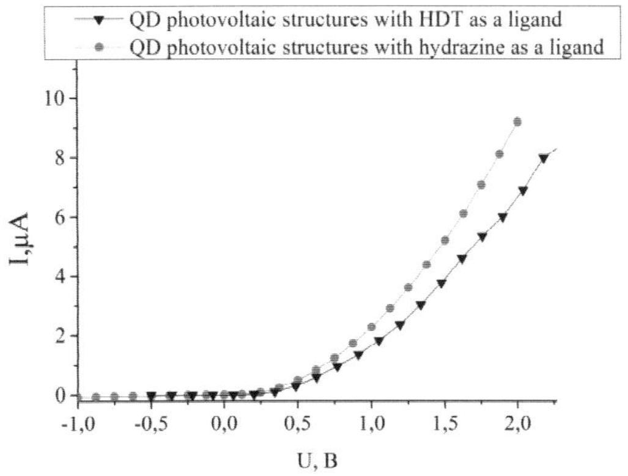

Figure 3. The current-voltage characteristics of a photovoltaic structure based on PbS QD film with different ligands in the dark.

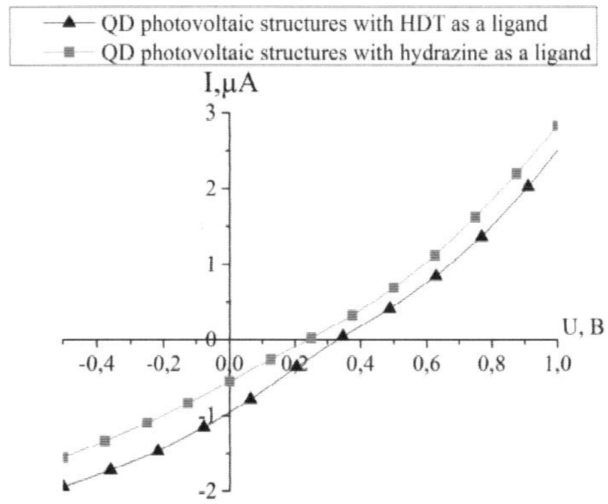

Figure 4. The current-voltage characteristics of a photovoltaic structure based on PbS QD film with different ligands under the halogen lamp irradiation.

4. Conclusion

Thus, the choice of surface ligands plays an important role in the efficiency of charge transfer between QDs in the film. We have shown that the decrease in the ligand length by a factor of 4 leads to a significant exponential increase in conductivity of the QD film, which is in a good agreement with the theory of the hopping mechanism of charge transfer in QD solids [JPCLett]. We propose that for the electrical conductivity of QD solids, the length of the ligand molecule is the crucial factor rather than its chemical structure. For PbS QD solid used as a photoactive layer in photovoltaic cells ITO/PEDOT: PSS/PbS/ZnO/Al, we found the optimal QD layer thickness of 70-90 nm which provides the enhanced performance.

Acknowledgments

The work was supported by "improving the competitiveness" program of the National Research Nuclear University «MEPhI». Preparation of QD solids was supported by the Russian Foundation for Basic Research, grant no. 16-29-09623 (OFI_M).

References

[1] Dayneko S, Linkov P, Martynov I et al 2016 *Physica-E: Low-dimensional Systems and Nanostructures* **79** 206
[2] Dayneko S, Tameev A, Tedoradze M et al 2013 *Appl. Phys. Lett.* **103** 063302
[3] Dayneko S, Tameev A, Tedoradze M et al 2014 *Proc. SPIE* **8981** 898113-1
[4] Fischer S A, Crotty A M, Kilina S V, Ivanov S A and Tretiak S 2012 *Nanoscale* **4** 904–914
[5] Kurian P A, Vijayan C, Suchand Sandeep C S, Philip R and Sathiyamoorthy K 2007 *Nanotechnology* **18** 75708
[6] Xu F, Gerlein L, Ma X, Haughn C, Doty M and Cloutier S 2015 *Materials* **8** 1858–1870

Resistance of magnetic field sensors based on molybdenum nanofilms to intense neutron fluxes

I Bolshakova[1], M Bulavin[2], N Kargin[3], Ya Kost[1], S Kulikov[2], M Radishevskiy[1], F Shurygin[1], M Strikhanov[3], I Vasil'evskii[3] and A Vasyliev[1]

[1]Magnetic Sensor Laboratory, Lviv Polytechnic National University
[2]Frank Laboratory of Neutron Physics, Joint Institute for Nuclear Research
[3]Institute of Functional Nuclear Electronics, National Research Nuclear University MEPhI

Corresponding author's e-mail address: inessa@mail.lviv.ua

Abstract. Influence of the reactor neutrons fluxes on parameters of Hall sensors based on molybdenum nanofilms had been investigated experimentally in real-time operation (on-line mode). Sensors sensitivity remained stable up to fluence of $1 \cdot 10^{24}$ n·m^{-2} at temperature $\approx$ 130 °C. It allows to recommend molybdenum sensors as the radiation-resistant magnetic diagnostic front-end components for fusion reactors type of ITER and DEMO.

1. Introduction

Modern R&D in the field of controlled thermonuclear fusion reached the conceptual design's level of demonstrational fusion reactors, which are called to confirm a viability of electrogenerating systems based on an energy utilization of neutrons and α-particles produced at the light nuclei coalescence in deuterium-tritium plasma. A flagship project in this direction is the European demonstrational reactor tokamak DEMO that is being developed by the EUROfusion Consortium [1]. A magnetic diagnostics' tools creation is the key task for effective and safety operation of the DEMO type reactors, which could help to resolve the humanity's energy problems and to reduce significantly the adverse impact on ecology already in the 21st century.

It is typical for the deuterium-tritium fusion reactors to have the high radiation (the neutrons, gamma-quanta) and thermal loads within a structure, which lead to characteristics degradation of functional materials used in diagnostic systems. Therefore, the metal inductance coils, which have relatively high tolerance to the radiation and thermal loads, are mainly used as the front-end components (sensors) of magnetic diagnostics. The main disadvantage is that the coils output signal is proportional not to the magnetic induction B, but to the rate of its change $\partial B/\partial t$ [2]. So, a coils usage requires the time integration of signal. This leads to errors accumulation due to a number of radiation-induced effects [3]. Nevertheless, the DEMO type reactors are planned to use basically in the steady-state mode when duration of the field pulses' flat-top parts ($B \approx$ const) will be reach hundreds of hours [3]. In this case pick-up coil based systems as reported in [4] become not so applicable. For operation under these conditions it is more appropriate to use the sensors based on Hall effect which can directly measure B with high precision in the case of both rapid-variable and steady-state magnetic fields.

Earlier the authors of this article have been developed and tested the semiconductor radiation-resistant Hall transducers [5]. In particular, it was confirmed that the sensors based on heterostructures of the In-containing III-V compounds are workable under environment of real fusion reactors (the JET tokamak, the testing was started on 2009 and continues up today). It was shown experimentally in the research nuclear reactors that they kept a functionality under neutron irradiation up to the limit fluence of $2 \cdot 10^{22}$ n·m^{-2} [5]. This corresponds to a radiation environment expected at the ex-vessel magnetic field sensors locations in the ITER experimental tokamak throughout its life time (~ 20 years) [6].

In the next generation fusion reactors of DEMO type the radiation fluxes and neutron fluences at the sensors locations will be hundreds and thousands of times higher than the ITER ones: in the ex-vessel area the fluences will reach $F \sim (10^{23} - 10^{24})$ n·m^{-2}, whereas in the in-vessel area it will be $F \sim (10^{25} - 10^{26})$ n·m^{-2}. The temperature environment for sensors in DEMO also will be much more rigid: (300 – 500) °C instead of (120 – 250) °C in ITER.

An analysis showed that Hall sensors based on the polycrystalline metal nanofilms can satisfy to these environment. As one of the reasons for their parameters stability under radiation loads it can be considered a presence of the branched network of grain boundaries in material, which are favor a recombination of the interstitial atoms and vacancies formed by the neutron bombardment [7].

At on-line testing in the IBR-2 reactor the authors of this article showed that sensors based on the polycrystalline gold nanofilms demonstrate a high radiation resistance up to fluence $F \approx 1 \cdot 10^{24}$ n·m^{-2} which was been accumulated for 1852 hours of reactor operation [8]. A character of the sensors sensitivity's dependence on the neutron fluence indicates that reached fluence level is not a limit one for these sensors.

In [9] Hall sensors based on Bi was proposed for ITER. But this metal has low melting temperature $T_m = 271$ °C, which allows to use it without additional cooling only ex-vessel where the thermal loads in ITER will be $T \approx 100$ °C, and this is close to the Bi recrystallization temperature $0.4 \cdot T_m \approx 108$ °C. At the same time in DEMO the working temperatures of the magnetic diagnostics' front-end components expected at the level $T \approx 200$ °C (ex-vessel, vacuum vessel) and $T \approx 300$ °C (in-vessel, blanket) [10]. This requires to use the materials with the higher melting temperature T_m than Bi has.

As a material for Hall sensors in the DEMO environment the molybdenum could be prospective which combines a high melting temperature $T_m \approx 2617$ °C with a small thermal neutron cross-section $\sigma \approx 2.6$ barn, which almost 40 times smaller than in Au. In addition, Mo has small linear expansion coefficient, and its Hall coefficient is almost independent on T in the range $T = (0 - 600)$ °C [11]. These factors should provide a high stability of Hall sensors based on Mo under the DEMO radiation and thermal loads.

The goal of this work is a radiation resistance testing of Hall sensors based on polycrystalline molybdenum nanofilms under irradiation by reactor neutrons up to the DEMO-relevant fluences.

2. Sensors and method of their on-line testing under neutron fluxes

Due to materials activation in consequence of irradiation by neutron fluxes the Hall sensors were tested in the on-line mode that intends a distant automatic parameters measurement with the help of special measuring systems as well as an obtained results transmission onto remote server.

2.1. Hall sensors

Investigated samples were manufactured on the basis of polycrystalline Mo nanofilms with a thickness of 50 nm, which were deposited onto Al_2O_3 (sapphire) substrate by the vacuum magnetron sputtering method. Ti-based contact metallization is used for both high temperature operation and low sheet resistance of the structure [12]. Identical four-leads Hall sensors in the form of symmetrical crosses with the (200 × 200) μm^2 active area's dimensions were formed using the lift-off process. The dimensions of separate chip were ≈ (1 × 1) mm^2 after dicing. Additional gold contact pads with a thickness of (1.5 – 2.0) μm was deposited onto sensors leads by the vacuum thermal evaporation method, and the ∅ 30 μm gold wires were jointed to them by the thermocompression bonding method to conduct further Hall measurements.

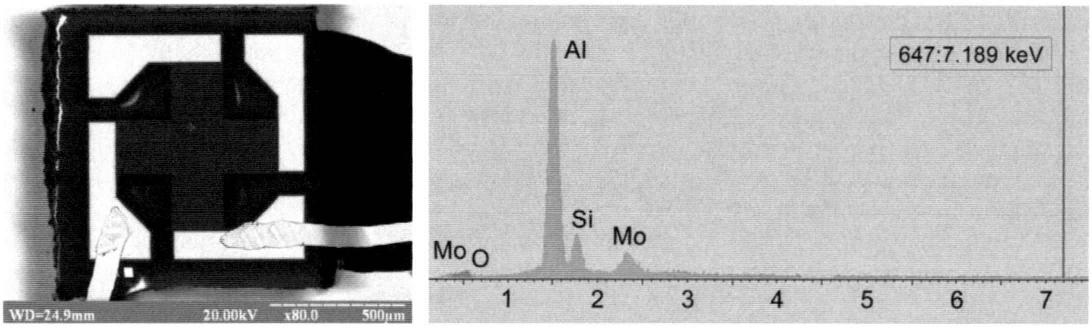

(a) (b)

Figure 1. The phase-contrast SEM image of the Mo-based Hall sensor with auxiliary wires for the charge removal (a), and the energy-dispersive X-ray spectrum for part of the sensor active area (b).

Figure 1 demonstrates the phase-contrast image of molybdenum Hall sensor (a) and the energy-dispersive X-ray spectrum for part of its active area's surface (b), which were obtained by the PEMMA-102-02 scanning electron microscope (SEM). The presence of Si spike on figure 1 b can be explained by the X-ray photons energy loss on the fluorescence in material of silicon detector.

2.2. Measurement of Hall sensors sensitivity

The testing of the Hall sensors radiation resistance in neutron fluxes were carried out by measuring in on-line mode the dependence of their sensitivity S on fluence F:

$$S(F) = \frac{1}{I} \cdot \frac{U_H(B, F)}{B}, \tag{1}$$

where S is the sensitivity, I – sensor bias current, B – magnetic induction, U_H – Hall voltage. Since the samples temperature T was fixed during on-line testing, its influence does not take into account in (1).

The S value was defined with taking into account the nonzero parasitic offset voltage U_0 at $B = 0$ that is adding to U_H, and, as a result, the sensor output signal is the superposition:

$$U(B, F) = U_H(B, F) + U_0(F), \tag{2}$$

where U is the output voltage, $U_0 = U(B = 0)$ – offset voltage.

For the metal sensors U_0 sometimes exceeds U_H because Hall signal in metals is a very low. Furthermore, the U_0 value can vary significantly for different samples. So, the key task is minimization of U_0 in order to rise a determination accuracy of true Hall voltage U_H. For that in this work it was used the spinning-current method that intends a quick successive circular synchronous interchanging of the current-leads pair and the voltage-leads pair of the sensor at $B = \text{const}$ and $F = \text{const}$ [13]. In so doing an averaging of four U values measured over one current spinning cycle with duration of t_{SC} gives the modified output signal that can be expressed as:

$$U^{SC}(B, F) = U_H(B, F) + U_0^{SC}(F), \tag{3}$$

where $U_0^{SC} << U_0$ is the remaining offset voltage. The spinning-current method also allows to reduce essentially a noises and interferences content in output signal, which is especially important at the using of long communication lines during testing in neutron fluxes.

The U_0^{SC} value can be very low. Modern versions of the spinning-current method allow to reduce the offset voltage by $U_0/U_0^{SC} \approx 10^4$ times [14]. Nevertheless, at the assessment of a Hall sensors radiation resistance it cannot be assumed that U_0^{SC} in (3) is equal to zero, because U_H and U_0^{SC} are depend on fluence F in different ways. Correspondingly for the more precise determination of $U_H(F)$ and $S(F)$ it needs to take into account the $U_0^{SC}(F)$ dependence.

For that purpose in this work the sensors were placed into the periodic pulsed test magnetic field:

$$B = \begin{cases} B_t, & k \cdot P \le t < (k+D) \cdot P, \\ 0, & (k+D) \cdot P \le t < (k+1) \cdot P, \quad k = 0,1,2,.. \end{cases} \tag{4}$$

where B_t is the test magnetic field amplitude, P – field pulses period, D – duty cycle. Demodulation of the output signal obtained in this way allow to separate two components:

$$\begin{aligned} U_1^{SC}(F) &= U^{SC}(B = B_t, F) = U_H(B_t, F) + U_0^{SC}(F), \\ U_2^{SC}(F) &= U^{SC}(B = 0, F) = U_0^{SC}(F), \end{aligned} \tag{5}$$

the difference of which allows to find the Hall voltage:

$$U_H(B_t, F) = U_1^{SC}(F) - U_2^{SC}(F). \tag{6}$$

During on-line testing the sensors bias current was $I = 40$ mA, the test magnetic field amplitude was $B_t = 15$ mT, the pulses period was $P = 4$ min., the duty cycle was $D = 0.5$. The one measurement's cycle duration of U^{SC} by the spinning-current method was $t_{SC} = 0.1$ sec., which is much less then duration of one field pulse $P \cdot D = 2$ min. This allowed to do a necessary number of measurements at $B = \text{const}$, and to avoid the transient processes that arise at a pulsed field's switching. The initial sensors sensitivity before an irradiation was $S_0 = S(F = 0) \approx 1.74$ mV·A^{-1}·T^{-1}.

2.3. The testing in neutron fluxes

The radiation resistivity testing of sensors based on the molybdenum nanofilms was conducted in the IBR-2 fast pulsed reactor in the irradiation channel #3 (JINR, Dubna), where the neutron energy spectrum is close to the forecasted one for the DEMO fusion reactor [15]. The general scheme of on-line irradiation experiment is presented on figure 2.

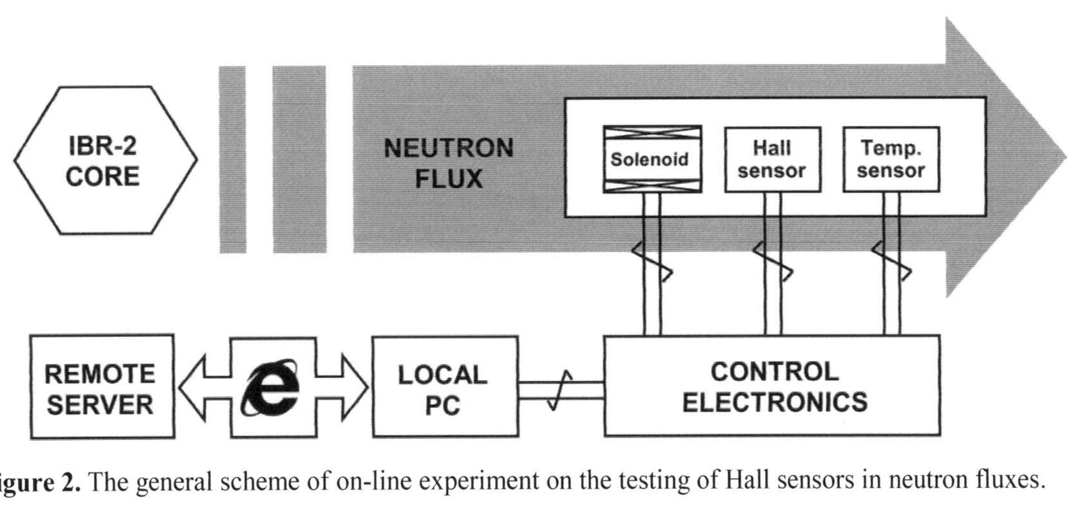

Figure 2. The general scheme of on-line experiment on the testing of Hall sensors in neutron fluxes.

The special fixture on the base of radiation- and thermal-resistant ceramics MACOR (Corning, USA) was developed for a samples placement in neutron flux, its elements are presented on figure 3 (a – c). For creation of test magnetic field (4) with B_t amplitude the ∅ 10 mm copper solenoid (figure 3 c) was used, inside which the sensors were disposed. The temperature was measured by the platinum resistance thermometer. The fixtures with sensors were mounted into the duralumin

IOP Conf. Series: Materials Science and Engineering **475** (2019) 012026 doi:10.1088/1757-899X/475/1/012026

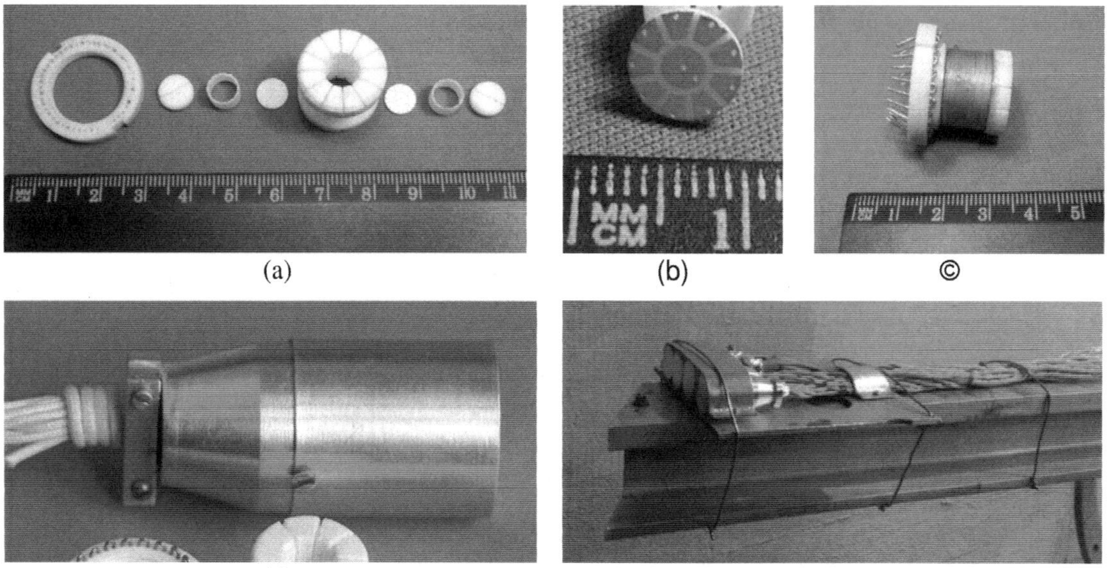

(a) (b) © (d) (e)

Figure 3. The fixture for a samples placement at the testing in neutron fluxes: the structural components (a); the plate for a sensors fixing (b); the solenoid for test magnetic field creation (c); the fixture with sensors inside the duralumin protective screen (d); the fixtures allocation on the supporting beam in the channel #3 of the IBR-2 reactor (e).

protective screens (figure 3 d) and then were fixed on the supporting beam in the irradiation channel (figure e).

For on-line measurements the special highly-sensitive control electronics was developed (figure 2) that realized an algorithm for measurement of small voltage U^{SC} based on the spinning-current method. This electronics provided the sensors bias current, controlled a solenoid operation, registered the temperature sensor readings, and transmitted obtained data to local PC. The specially designed software performed the $U^{SC}(F)$ demodulation, separated the $U_1^{SC}(F)$ and $U_2^{SC}(F)$ components, and provide their following digital filtering. On the basis of this data it was computed the $U_H(B = B_t, F)$ dependence adjusted for $U_0^{SC}(F)$, as described in subsection 2.2. Obtained results were transmitted to remote server via Internet for the subsequent calculation of *S*(*F*).

Due to intensive neutron flux the control electronics and local PC were placed outside the irradiation channel – in the IBR-2 reactor hall at a distance of $\approx$ 15 m and in the control room at a distance of $\approx$ 40 m correspondingly. The measuring system components were interconnected by the radiation-resistant cables of the "twisted pair" type (figure 2) in order to reduce an influence of external interferences on measured signals.

During experiment the neutron flux intensity at the location of fixtures with sensors was $\approx 1.5\times10^{17}$ n·m^{-2}·s^{-1}, which corresponds to the expected in-vessel radiation environment (blanket) in the DEMO reactor [16]. The total irradiation time was 1852 hours, which was allowed to obtain the DEMO-relevant maximum fluence $F_m \approx 1\times10^{24}$ n·m^{-2}.

The samples temperature during experiment was $T = (130 \pm 5)$ °C.

3. Results and discussion

In the result of conducted on-line testing it was obtained about of 45 thousands data points. Figure 4 demonstrates the random sampling from the measured *S*(*F*) dependence. As can be seen the molybdenum sensors in general preserve the sensitivity's stability in the whole considered fluence

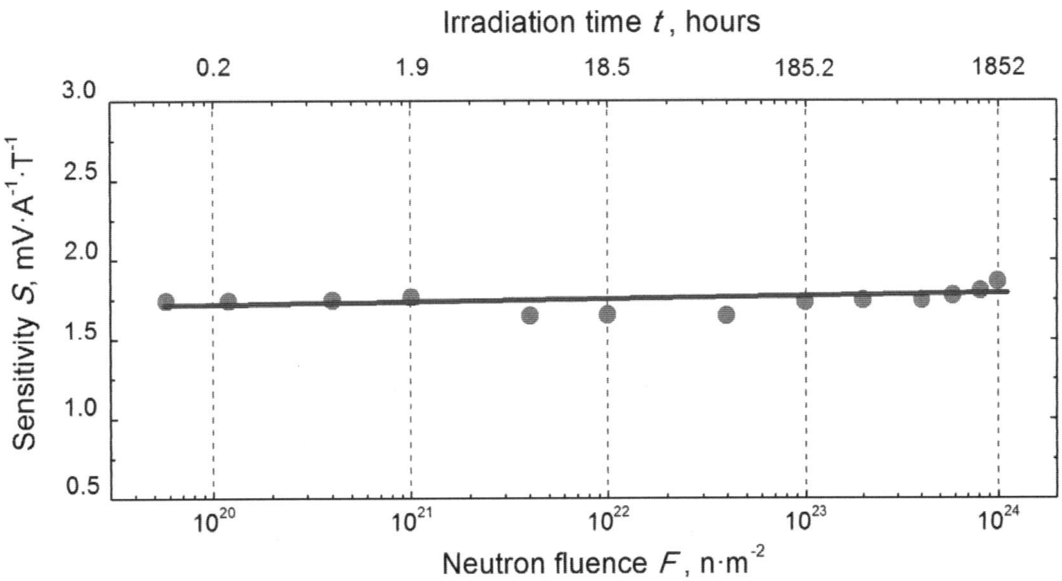

Figure 4. The sensitivity dependence on neutron fluence $S(F)$ (bottom axis) and on irradiation time $S(t)$ (top axis) for the Hall sensors based on polycrystalline Mo nanofilms: circles mark the measured data, line is the linear approximation.

interval up to maximum value $F_m \approx 1{\times}10^{24}$ n·m^{-2}. By the end of experiment the relative deviation of S form the initial value, $|S(F_m)-S_0|/S_0$, did not exceed 1 % for all investigated samples. The possible explanation of such high radiation resistance is a presence in the polycrystalline nanofilms of the grain boundaries network, which are favor a recombination of the interstitials and vacancies created under neutron bombardment [7].

The data points rejections from the $S(F) = S_0$ level on figure 4 can be explained by the operation peculiarities of the IBR-2 reactor (the irradiation sessions are continue 12-15 days with the interruptions form 1 till several weeks) as well as by the insignificant deviations of its output power. This led to the samples temperature changes within ±5 °C. Although such T variations do not affect the Hall voltage U_H of investigated samples (which is also confirmed by data for Mo obtained in [11]) but they can change the offset voltage U_0^{SC}. At the same time, it should be noted that the spinning-current method does not remove the offset completely from the measured signal, and consequently the dependence of U_0^{SC} on T may reflected to some extent on the results of U_H measurement. A big problem is the small U_H value typical for most of metals. As a result at the on-line investigations with the using of long communication lines the electrical noises and interferences become the source of additional inaccuracies of measurement.

An overcoming of these obstacles requires to improve equipment for the metal Hall sensors on-line investigations, in particular that is based on advanced versions of the spinning-current method [14]. Moreover, the improvement of the molybdenum sensors manufacturing technologies is needed to reduce the offset voltage and own noises level, to improve a reproducibility of devices characteristics etc.

4. Conclusions

The sensitivity S of the Hall sensors based on the polycrystalline Mo nanofilms is stable under an action of reactor neutrons fluxes up to the fluence $F \approx 1{\times}10^{24}$ n·m^{-2} at the temperature $T \approx 130$ °C. This confirms the possibility of their using as the ex-vessel front-end components of magnetic diagnostics for the DEMO type's fusion reactors where the radiation loads $F \sim (10^{23} - 10^{24})$ n·m^{-2} are forecasted. The absence on $S(F)$ dependence of signs of the Mo sensors parameters degradation in the considered fluences F range allows to expect that the sensors sensitivity will not change significantly at the higher fluences up to $F \sim (10^{25} - 10^{26})$ n·m^{-2} typical for the in-vessel magnetic field sensors' positions in DEMO. Nevertheless, the testing continuation is needed to experimentally confirm the Mo sensors radiation resistance in such environment. Moreover, the assessment of their thermal stability in the DEMO's temperatures range is also needed.

Acknowledgments
The work had been carried out with the support of the Ministry of Education and Science of Ukraine (DB/RAD Project), the EUROfusion Consortium (WPDC Project), and the Competitiveness Program of the National Research Nuclear University MEPhI. The authors are thankful to Mr. Roman Serkiz (Ivan Franko Lviv National University) for the SEM investigations of samples as well as to Dr. Vincent Mosser (Issy Technology Center, Itron France) for the useful consultations as for the spinning-current method.

References
[1] Federici G et al 2016 *Fusion Eng. Design* **109-111** 1464
[2] Hutchinson I H 2002 *Principles of Plasma Diagnostics* (Cambridge: University Press)
[3] Orsitto F S et al 2016 *Nucl. Fusion* **56** 026009
[4] Ma Y, Vayakis G, Begrambekov L B et al 2016 *Fusion Eng. Des.* **112** 594
[5] Bolshakova I et al 2015 *Nucl. Fusion* **55** 083006
[6] Loughlin M J 2008 IDM number iter_d_2f6s7y
[7] Ackland G 2010 *Science* **327** 1587
[8] Bolshakova I et al 2017 *Nucl. Fusion* **57** (*In press*, DOI: 10.1088/1741-4326/aa7867)
[9] Kocan M et al 2017 *Fusion Eng. Des.* (*In press*, DOI: 10.1016/j.fusengdes.2017.03.043)
[10] Koncara B, Draksler M, Costa Garrido O and Meszaros B 2017 *Fusion Eng. Des.* (*In press*, DOI: 10.1016/j.fusengdes.2017.03.134)
[11] Frank V 1959 *Appl. Sci. Res.* B**7** 41
[12] Shostachenko S A, Zakharchenko R V, Ryzhuk R V et al 2016 *Proc. SPIE, International Conference on Micro- and Nano-Electronics* (doi:10.1117/12.2267053)
[13] Popovic R S 2004 *Hall effect devices* (Bristol and Philadelphia: IOP Publishing)
[14] Mosser V, Matringe N and Haddab Y 2017 *IEEE T. Instrum. Meas.* **66** 637
[15] Shabalin E P, Verkhoglyadov A E, Bulavin M V et al 2015 *Phys. Part. Nuclei. Lett.* **12** 336
[16] Gilbert M R, Eade T, Bachmann C, Fischer U and Taylor N P 2017 *Nucl. Fusion* **57** 046015

IOP Conf. Series: Materials Science and Engineering **475** (2019) 012027 doi:10.1088/1757-899X/475/1/012027

Quasi-two-dimensional electron-hole liquid with two type of holes in SiGe/Si quantum wells

A A Vasilchenko[1] **and G F Kopytov**[1]

[1]Kuban State University, Stavropolskaya 149, Krasnodar, 350040, Russia

Corresponding author's e-mail address: a_vas2002@mail.ru

Abstract. The density functional theory was used to calculate the energy of a quasi-two-dimensional electron-hole liquid (EHL) in $Si/Si_{1-x}Ge_x/Si$ quantum wells. We have performed a numerical solution of the three one-dimensional nonlinear Schrödinger equations for electrons, heavy and light holes. It is shown that EHL contains heavy and light holes for small x. With an increase in x, a transition to a state with one type of hole occurs, and the density of electron-hole pairs decreases by more than three times. The results of the calculations are in good agreement with the experimental results.

1. Introduction

Investigation of the properties of electron-hole systems is one of the actively developing areas of modern condensed matter physics. At present, the greatest attention is paid to low-dimensional electron-hole systems, in which the role of the Coulomb interaction increases. Theoretical and experimental work in this direction is mainly devoted to systems with a finite number of particles (excitons [1,2], trions [3,4], and electron-hole complexes with a number of particles less than 10 [5,6]).

The properties of a quasi-two-dimensional electron-hole liquid (EHL) in semiconductor structures are less studied. Recently, EHL has been detected in $SiO_2/Si/$ SiO_2 quantum wells [7,8] and in $Si/Si_{1-x}Ge_x/$ Si heterostructures [9-14] and GaAs/AlAs superlattices [15]. A fine structure in the photoluminescence spectra of the EHL in SiGe quantum wells with a small amount of Ge (several percent) both in the visible and in the infrared region was experimentally observed in [14]. It was possible to explain the shape of the condensed phase line within the framework of the concepts of a multicomponent Fermi liquid containing both heavy and light holes.

In this paper we use the density functional theory for finding the energy and the equilibrium density of a quasi-two-dimensional EHL with two types of holes. Three one-dimensional nonlinear Schrödinger equations are solved numerically for electrons, light and heavy holes.

2. Theoretical model

For quasi-two-dimensional EHL with electrons, heavy and light holes, the total energy is written as follows:

$$E_t[n_e,n_h]=T_e[n_e]+T_h[n_{hh}]+T[n_{hl}]+\frac{1}{2}\int V_c(z)(n_e(z)-n_h(z))dz+$$
$$\int U_e(z)n_e(z)dz+\int U_{hh}(z)n_{hh}(z)dz+\int U_{hl}(z)n_{hl}(z)dz+E_{xc}[n_e,n_{hh},n_{hl}], \quad (1)$$

where T_e , T_{hh}, T_{hl} are kinetic energy of electrons, heavy and light holes, respectively, $V_c(z)$ is the Coulomb potential, E_{xc} is the exchange-correlation energy, $U_e(z)$, $U_{hh}(z)$, $U_{hl}(z)$ are external potentials for electrons, heavy and light holes, respectively, n_e , n_{hh}, n_{hl} are density electron, heavy and light hole, respectively, $n_h = n_{hh} + n_{hl}$.

The exciton system of units is used: the energy is measured in units of $Ry_{ex}=e^2/2ka_{ex}$, and the length in units $a_{ex} = k\hbar^2 / \mu e^2$, where □ is the reduced mass, k is the dielectric constant.

Varying expression (1) over the densities n_e, n_{hh} and n_{hl}, we obtain three Schrödinger equations:

$$\left(-\frac{\mu}{m_{z,i}}\frac{d^2}{dz^2}+V_{eff,i}(z)\right)\psi_{n,i}(z)=E_{n,i}\psi_{n,i}(z), \quad (2)$$

where $i=e,\ hh,\ hl$.

Thus, the problem reduces to the solution of three one-dimensional nonlinear Schrödinger equations with potentials $V_{eff.e}(z) = V_c(z) + V_{xc.e}(z) + U_e(z)$, $V_{eff.hh}(z) = - V_c(z) + V_{xc.hh}(z) + U_{hh}(z)$ and $V_{eff.hl}(z) = - V_c(z) + V_{xc.hl}(z) + U_{hl}(z)$, where $V_{xc}(z)$ is the exchange-correlation potential, and the electrostatic potential is obtained from the Poisson equation:

$$V_c(z)=8\pi\int_{-\infty}^{z}(z-z')[n_h(z')-n_e(z')]dz', \quad (3)$$

The external potentials for electrons and holes are given by the expression:

$$U_i(z)=\begin{cases}0, |z|\le d/2\\ V_i, |z|>d/2\end{cases}, \quad (4)$$

where d is the width of the well (barrier).

With this choice of external potentials, the energy is measured from the width of the SiGe band gap. The Si/SiGe transition is a type-II heterostructure, for which $V_e < 0$, $V_{hh} > 0$ and $V_{hl} > 0$.
Due to deformations, the splitting of hole states occurs in the SiGe layer. The splitting of the valence band can be changed by applying a compensating pressure to the heterostructure. Below, the splitting between the heavy and light holes equals to zero.

When only the lower level of the size quantization is filled, the carrier densities are given by:

$$n_e(z)=N_e\psi_{0,e}^2(z)\ ,\quad n_{hh}(z)=N_{hh}\psi_{0,hh}^2(z)\ ,\ n_{hl}(z)=N_{hl}\psi_{0,hl}^2(z)\ , \quad (5)$$

where N_e, N_{hh}, N_{hl} are two-dimensional electron, heavy and light hole densities, respectively.
Below, the index "0" at variables is omitted.

The kinetic energy has the following form:

$$T_i[n_i]=\frac{\pi\mu}{g_i m_{d,i}}N_i^2+N_i(E_i-\int V_{eff,i}(z)\psi_i^2(z)dz). \quad (6)$$

In Eq. (6), the first term corresponds to the total kinetic energy of the carriers along the electron-hole layer, and the second corresponds to the kinetic energy across the layer.
For the exchange-correlation energy, we use the local density approximation:

$$E_{xc}[n_e,n_{hh},n_{hl}]=\int dz e_{xc}(n_e,n_{hh},n_{hl})\ , \quad (7)$$

where $e(n_e,n_{hh},n_{hl})$ is the exchange-correlation energy of electrons and holes per unit volume.

Thus, the exchange-correlation potentials have the form:

$$V_{xc,e}(z) = \frac{de_{xc}}{dn_e}, \quad V_{xc,hh}(z) = \frac{de_{xc}}{dn_{hh}}, \quad V_{xc,hl}(z) = \frac{de_{xc}}{dn_{hl}}. \tag{8}$$

In general case, the form of the expression for e_{xc} is unknown. In the case of neutral EHL $n_e(z) = n_{hh}(z)+n_{hl}(z) = n(z)$ the approximation formula is available for the exchange-correlation energy $\varepsilon_{xc} = e_{xc}/n$ [16]:

$$\varepsilon_{xc} = \frac{1}{2}\frac{a+br_s}{c+dr_s+r_s^2}, \tag{9}$$

where $r_s = \left(\frac{3}{4\pi n}\right)^{1/3}$, $a = -4{,}8316$, $b = -5{,}0879$, $c = 0{,}0152$, $d = 3{,}0426$.

We consider an electrically neutral EHL, then $N_e = N_{hh} + N_{hl}$. Taking into account the Fermi energy of charge carriers with the density-of-states mass density $m_{d,i} = (m_{x,i}m_{y,i})^{1/2}$ and the number of equivalent valleys g_i is equal to $E_{F,i} = 2\pi\mu N_i / g_i m_{d,i}$ we have

$$E_{hh} + \frac{2\pi\mu N_{hh}}{g_{hh}m_{d,hh}} = E_{hl} + \frac{2\pi\mu N_{hl}}{g_{hl}m_{d,hl}}. \tag{10}$$

From Eq. (10) with $g_i = 1$ for an electrically neutral EHL we obtain:

$$N_{hl} = N_e \frac{m_{d,hl}}{m_{d,hh}+m_{d,hl}} - \frac{m_{d,hh}m_{d,hl}}{m_{d,hh}+m_{d,hl}} \frac{E_{hl}-E_{hh}}{2\pi\mu}. \tag{11}$$

In case the expression (11) produces $N_{hl} < 0$, we take $N_{hh} = N_e$ u $N_{hl} = 0$.

Below, energy per electron-hole pair is measured from the energy level E_{hh} of heavy holes in the quantum well: $E_{eh} = -E_{hh} + E_t/N_e$.

3. Results

The system of nonlinear equations (2) - (11) was solved numerically. Calculations were carried out for the Si/SiGe/Si structure. For calculations, we used the following parameters [17, 18]: $V_e = 7{,}7x$, $V_{hh} = V_{hl} = 53{,}7x$, $g_e = 4$, $g_{hh} = g_{hl} = 1$, $m_{z,e} = 0{,}198m_0$, $m_{d,e} = 0{,}44m_0$, $m_{z,hh} = 0{,}28m_0$, $m_{d,hh} = 0{,}2m_0$, $m_{z,hl} = 0{,}18m_0$, $m_{d,hl} = 0{,}25m_0$, (m_0 is free electron mass). The calculations were carried out for a structure with a width of the well $d = 1$. The reduced mass has been taking from silicon $\mu = 0{,}126$, then $E_{ex} \approx 13\text{meV}$ and $a_{ex} \approx 5nm$.

The results of a numerical solution of the Schrödinger equations are shown in Fig. 1. The barrier causes the electron wave function to have a more extended state in comparison with the wave functions of the holes. Thus, holes create an attractive potential for electrons, which is more than twice the height of the barrier. Note that the exchange-correlation interaction also significantly reduces the height of the barrier for electrons.

Fig. 2 shows the dependence of the energy per one electron-hole pair on the two-dimensional density of pairs for different values of x. At $x = 0.03$, the equilibrium density of electron-hole pairs equal to 0.42 (1.7 10^{12} cm^{-2}), which is in a very good agreement with the experimental value [12,14]. With increasing x, the equilibrium density decreases and at x=0.09 it reaches the value of 10^{12} cm^{-2}. In this range EHL contains both heavy and light holes. Potential broadening at the heterointerfaces can be accounted as in [19], but for the wide enough quantum well this correction is not of crucial effect on the electron-hole pair energy.

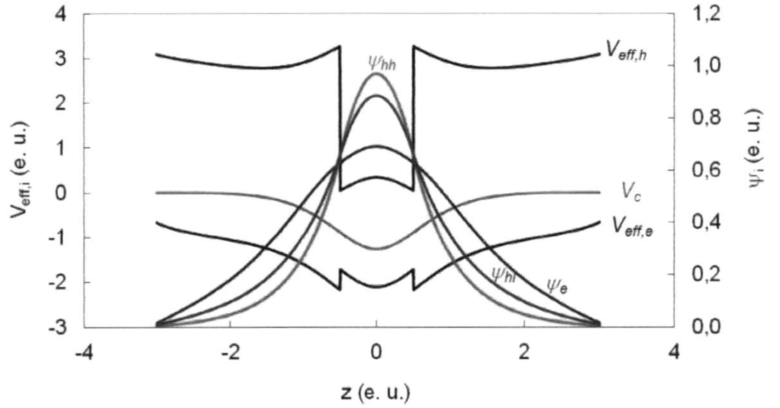

Figure 1. Profiles of wave functions and potentials for electrons, light and heavy holes ($N_e = 0.36$, $x = 0.06$).

The formation of EHL is possible if it is stable with respect to the decay into free excitons, that is, when the energy of the electron-hole pair, equal to - E_{eh}, is greater than the binding energy of the exciton. It was shown in [17, 20] that for $d > 1$ the binding energy of an exciton is close to the binding energy of a three-dimensional exciton. From the results presented in Fig. 2, it is seen that EHL binding energy is always over than 5 meV. The calculated binding energy is in good agreement with the experimental results [12, 14]. We note that the calculations for an EHL with heavy holes [21] give a much lower equilibrium density and binding energy.

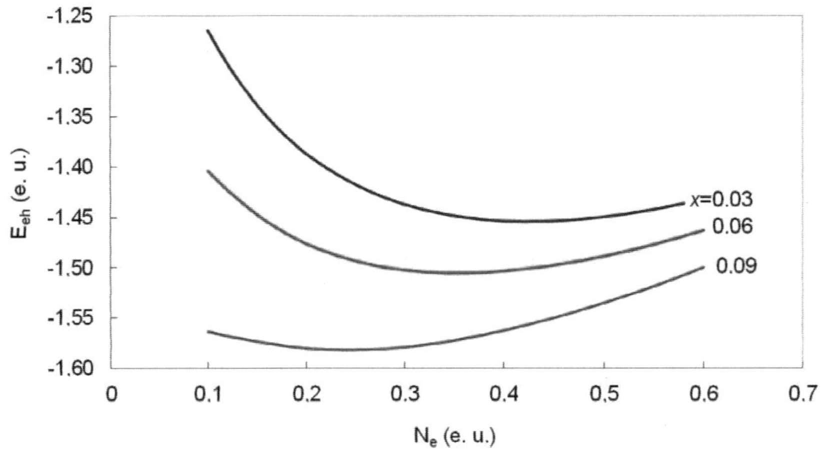

Figure 2. Dependence of the energy of an electron-hole pair on the two-dimensional density of electron-hole pairs.

At $x > 0.1$, there are only heavy holes in EHL (Fig. 3). This is due to the fact that as the well depth increases, the difference between the energy levels of light and heavy holes increases, and the potential energy of light holes outside the quantum well increases. EHL with heavy holes has a considerably lower equilibrium density than EHL with two types of holes. The decrease of equilibrium density with increasing x is mainly due to the influence of the longitudinal kinetic energy, which is smaller in the SiGe layer for light holes.

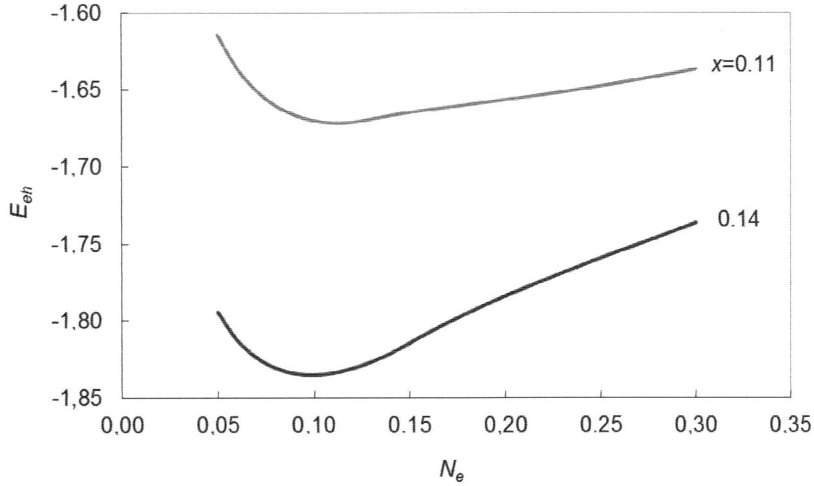

Figure 3. Dependence of the energy of an electron-hole pair on the two-dimensional density of electron-hole pairs.

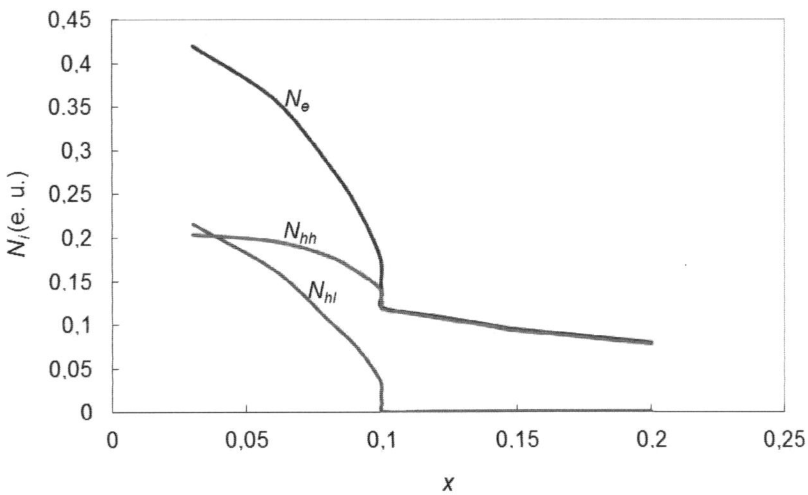

Figure 4. Dependence of two-dimensional densities of charge carriers on x.

Fig. 4 shows the dependence of the two-dimensional densities of charge carriers on Ge concentrations in the SiGe layer. It can be seen that for small x the densities of light and heavy holes are close. At $x > 0.1$ the EHL contains only heavy holes. Note that this transition occurs in structures with splitting of hole states at a much smaller value of x [22].

4. Conclusion

In this paper, we have calculated the EHL energy in the Si/$Si_{1-x}Ge_x$/ Si quantum wells. It is shown that at low germanium concentrations, the EHL contains two types of holes The minimum energy corresponds to the equilibrium density of electron-hole pairs, for instance, at $x = 0.03$ the equilibrium density $N = 1.7\ 10^{12}\ \text{cm}^{-2}$. The calculated density agrees very well with the experimental results. With increasing x, a transition to a state with one type of hole occurs, and in this transition the density of electron-hole pairs decreases by more than three times. For $x > 0.1$, the EHL contains only heavy holes. Note that such a transition was observed in the experiment.

Acknowledgments

The work was supported by the Russian Foundation for Basic Research and by the Ministry of Education and Science of the Krasnodar Region (Project No. 16-42-230280).

References

[1] Sidor Y, Partoens B, Peeters F M et al 2006 *Phys. Rev.* B **73** 155334
[2] Butov L V, Gossard A C and Chemla D S 2002 *Nature* **418** 751
[3] Bracker A S, Stinaff E A, Gammon D et al 2005 *Phys. Rev.* B **72** 035332
[4] Anisimovas E and Peeters F M 2003 *Phys. Rev.* B **68** 115310
[5] Vasilchenko A A and Tolkachev D M 2016 *Mod. Phys. Lett.* B **30** 1650013
[6] Kärkkäinen K, Koskinen M, Manninen M and Reimann S M 2004 *Sol. St. Comm.* **130** 187
[7] Pauc N, Calvo V, Eymery J, Fournel F and Magnea N 2004 *Phys. Rev. Lett.* **92** 236802
[8] Pauc N, Calvo V, Eymery J, Fournel F and Magnea N 2005 *Phys. Rev.* B **72** 205325
[9] Zaitsev V V, Bagaev V S, Burbaev T M et al 2008 *Physica* E **40** 1172
[10] Bagaev V S, Zaitsev V V, Burbaev T M et al 2008 *Zh. Exp. Teor. Fiz.* **134** 988
[11] Bagaev V S, Krivobok V S, Nikolaev S N et al 2010 *Phys. Rev.* B **82** 115313
[12] Bagaev V S, Krivobok V S, Nikolaev S N et al 2011 *Pisma Zh. Exp. Teor. Fiz.* **94** 63
[13] Bagaev V S, Krivobok V S, Nikolaev S N et al 2013 *Zh. Exp. Teor. Fiz.* **144**, 1045
[14] Nikolayev S N, Krivobok V S, Bagayev V S et al 2016 *Pisma Zh. Exp. Teor. Fiz.* **104** 161
[15] Furukawa Y and Nakayama M 2015 *Journal of Physics: Conference Series* **619** 012005
[16] Vashishta P and Kalia R K 1982 *Phys. Rev.* B, **10**, 6492
[17] Penn C, Schäffler F, Bauer G and Glutsch S 1999 *Phys. Rev.* B **59** 13314
[18] Yang L, Watling J R, Wilkins R C W et al 2004 *Semicond. Sci. Technol.* **19** 1174
[19] Sibirmovsky Yu D, Vasil'evskii I S and Kargin N I 2016 *J. of Phys.: Conf. Series* **690(1)** 012016
[20] Bagaev V S, Krivobok V S, Martovitskii V P and Novikov A V 2009 *Zh. Exp. Teor. Fiz.* **136** 1154
[21] Vasilchenko A A, Kopytov G F, Krivobok V S and Ermokhin D A 2017 *Russian Physics Journal* **59** 1693
[22] Vasilchenko A A and Kopytov G F 2017 *Proc.VIII Int. Conf. on Physics and Technology of Nanoheterostructure Microwave Electronics (In honour of RAS Corresponding Member, Professor V G Mokerov)* (Moscow: National Research Nuclear University «MEPhI» Press) p12

IOP Conf. Series: Materials Science and Engineering **475** (2019) 012028 doi:10.1088/1757-899X/475/1/012028

Influence of the length of organic molecules of ligands on the PbS QD solids optical properties

Y V Goltyapin[1], M A Zvaigzne[1], A A Chistyakov[1], V R Nikitenko[1], P S Samokhvalov[1], I L Martynov[1] and A R Tameev[2]

[1]National Research Nuclear University MEPhI (Moscow Engineering Physics Institute), Moscow, 115409, Russia
[2]A.N. Frumkin Institute of Physical Chemistry and Electrochemistry of the Russian Academy of Sciences, 31-bld.4 Leninsky Prospect, Moscow 119071, Russia

Corresponding author's e-mail address: goltyav@gmail.com

Abstract. The quantum dot (QD) solids are promising materials for use in electronics. We studied the effect of variations in the length of organic ligands on the optical properties of PbS QD solids from the measurements of photoluminescence spectra and decay kinetics. The decay is becoming exponentially faster along with the decrease of the length of ligand molecules, which is consistent with the decay of excitons by hopping of charge carriers.

1. Introduction

In recent years, the great attention is devoted to the optical and electrical phenomena in the so-called QD-solids. This term means a new type of a condensed matter, which is formed by "artificial atoms" – metallic or semiconducting nanocrystals. QD solids are very promising for electronics, in particular related to photovoltaics [1-4] and electroluminescence [5,6], because one can manage the optical and electronic properties due to quantum confinement effect. Charge transport is an important process for the operation of electronic devices [7,8]. Although great activity was developed in order to obtain well-ordered structures (superlattices), positional and hence energetic disorder of electronic (hole) states remain a very important factor, in particular, due to scatter in a size of QDs [7]. Another important source is a disorder in a non-ideal structure of shell around every QD, which is formed by surface-active organic molecules (ligands). These shells are considered typically as spacers, separating different QDs, hence supporting the quantum confinement, but the function of ligands in QD solids is not completely clear [9,10]. The disorder in combination with the weakness of electronic coupling of different QDs can produce a localization of all electronic (hole) states in a QD-solid [11], reminiscent to organic semiconductors [2,8]. The nature of transport in semiconductor QD solids is still under debates. It was considered as band-like process via extended states [1], while the hopping nature of transport also was supported [2,7,8]. The latter case seems to be very probable in a rather disordered QD-solids [2,7]. In this work, we replaced ligands in a PbS QD-film and analyzed the influence of variations in the length of molecules of ligands on the photoluminescence spectra and decay kinetics.

2. Experimental

In this work, QDs PbS synthesized by the colloidal chemistry method were used, and they have a typical diameter of 4 nm. The surface of QDs was covered with a monolayer of molecules of oleic acid (OA) as a result of the synthesis. The initial ligands were replaced by new ligands, such as hydrazine, and 1.6-hexanedithiol (HDT). The lengths of new surface molecules, respectively, are 4 and 2 times

smaller in comparison with the OA molecule.
Hydrazine is widely used to replace long-chain ligand molecules on the surface of PbS QDs. In turn, HDT is still poorly researched and is used for the first time, to our best knowledge, as a surface molecule for the formation of condensates of quantum dots. The presence of a thiol group in the composition of the HDT suggests that it will be used in the "role" of the ligand for QDs PbS, due to the high affinity for sulfur molecules, which allows chemically to "sew" the QDs in the film.
QD-solids with the initial ligands were thin films obtained by the spin-coating method. The replacement for new ligands was carried out directly in the process of the film preparation. A solution of QD (with OA as a ligand) in toluene with a volume of 50 μl and a concentration of 10 mg / ml was applied to a pre-cleaned glass substrate, which was then rotated for 30 seconds at an angular rate of 1500 rpm in a spin-coating setup. After that, 50 μl of the solution of the molecules of the new ligand (for hydrazine, 1 mole / L in methanol, for HDT 0.1 g / liter in acetonitrile) was applied to the resulting film using an automatic pipette, and after 30 seconds of exposure the film was rotated 30 seconds with an angular speed of 1500 rpm. Further, to remove excess molecules of ligands from the surface of the film, a suitable solvent for each ligand was applied to it, and the film was rotated for 30 seconds at a speed of 1500 rpm again. The described procedure was repeated 5 times to obtain the required thickness. The thickness of the obtained samples, 40-100 nm, was measured with a microinterferometer.
Control of replacement of surface molecules was carried out by analyzing the IR absorption spectra, obtained with the Varian IR spectrophotometer "Scimitar1000 FT-IR" for QD films.
For all samples of the films and the QDs solution, the luminescence and absorption spectra were measured, as well as the luminescence decay kinetics. For research of the absorption spectra, we used the halogen lamp Ocean Optics LS-1 for a wide spectral range, a semiconductor laser with a radiation wavelength of 908 nm was used to excite the luminescence. The spectra were recorded with the Avantes AvaSpec-NIR256-1.7 spectrometer.
For the research of the kinetics of luminescence, the second harmonic of the picosecond YAG: Nd3 + laser was used (532 nm wavelength, pulse duration 350 ps, pulse repetition frequency 50 Hz, average power 3 mW). The signal was recorded at the wavelengths of the luminescence maximums with the help of a high-speed PIN photodiode.

3. Results

As follows from Figure 1, a red shift of the luminescence spectra of the QD films was observed in the respect to the solution, which became more pronounced when the ligand was changed to a shorter one. The position of the first exciton absorption maximum remained unchanged after the transition from the solution to the condensed state. In addition, the width of the luminescence spectra varied. If the transition from a solution of QD with OA as a ligand to a film occurs, the spectrum was broadened and amounted to $\Delta\lambda = 195 \pm 8$ nm. If the original ligands were replaced with HDT in the film, the luminescence spectrum narrowed to $\Delta\lambda = 159 \pm 8$ nm, and if it was replaced by hydrazine, it significantly broadened and amounted to $\Delta\lambda = 320 \pm 8$ nm.

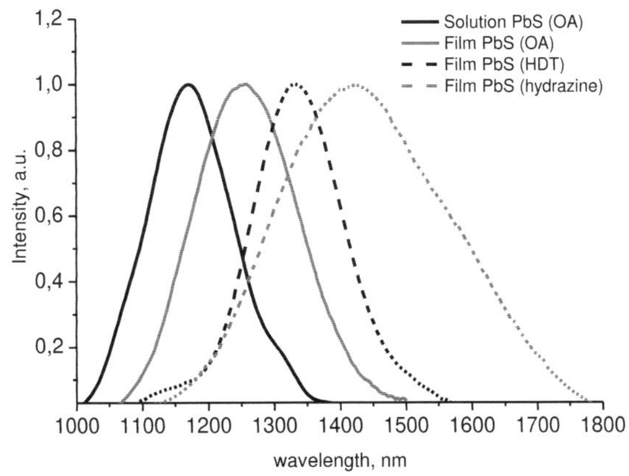

Figure 1. FL spectra of a solution and QD films with different surface ligands.

When measuring the kinetics of luminescence decay, the typical biexponential time dependence of decay kinetics [9] was observed in QD-films with initial ligands (OA):

$$I(t)=A_1e^{\left(-\frac{t}{\tau_1}\right)} + A_2e^{\left(-\frac{t}{\tau_2}\right)}, \quad (1)$$

where τ_1 and τ_2 represent the fast-decay and the slow-decay component, respectively; A_1 and A_2 are their amplitudes. We obtained $\tau_1 = 40$ ns and $\tau_2 = 160$ ns, while in a solution decay kinetics is monoexponential with the decay time $\tau_0 = 570$ ns, see Figure 2.

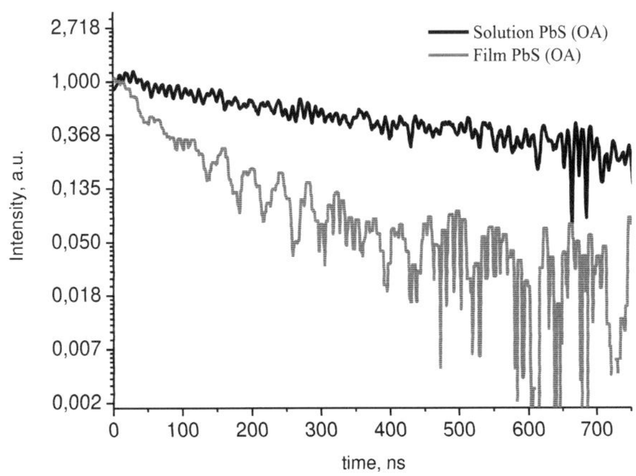

Figure 2. Kinetics of the luminescence decay of a solution and QD films with initial ligands.

A significant decrease in the decay time of luminescence was observed in the films with replaced ligands, namely 15 and 7 ns for the case of ligands HDT and hydrazine, respectively. Probably, monoexponential kinetics is observed.

4. Discussion

The width of the luminescence spectrum is directly related to the spreading of QDs over the size inside the film. The spreading could be apparently increased by an aggregation of QDs. One can assume that an aggregation is sufficient after replacing HDT ligands with hydrazine, reflected on the increased width of the spectrum, see the Figure 1, while in a QD film with a HDT, the particles are fairly uniformly distributed over the sample.

The observed redshift in the photoluminescence peaks of QDs, see the Figure 1, can be mainly due to four possible causes [12]:

a) increase in the size of QDs, that is not occurs in this case;
b) stronger electronic coupling between neighboring QDs with shorter ligands, which reduces the quantum confinement and increases the probability of charge transfer over the manifold of size-distributed QDs; an electon (a hole) is transferred most probably from a nanoparticle with a wider energy gap to a nanoparticle with a smaller energy gap at initial time interval;
c) the Förster-type resonant transfer of an exciton; this process of energy transfer will lead to the formation of an asymmetric spectrum due to the extinction of photoluminescence from the high-energy side [12]; however, the symmetry of the luminescence spectra is observed from Figure 1, precluding the exciton transfer;
d) charge transfer between QD and surface ligands; one has to note, that the charge transfer between QD and the surface ligand is not a noticeable process in the PbS QD films [13].

Therefore, one can assume that a stronger coupling of electrons and respective facilitation of charged transfer to neighbor QDs is the main factor, contributing to the red shift, observed in the luminescence of QD films with replaced ligands. This fact allows one to assume the appearance of an additional channel of nonradiative relaxation after replacing of initial ligands to the shorter ones. An analysis of the possible mechanisms of nonradiative processes under the conditions of our experiment [14,15] suggests that this is either resonant energy transfer between QDs of various sizes (excluded, as was mentioned above), or transfer of charge carriers between QDs. The transfer rate increases exponentially along with the decrease of a length of a ligand molecule, leading to the sharp increase of a photocurrent [16]. Despite the aggregation of QDs and the non-uniform spatial distribution of particles inside the film with hydrazine, the luminescence kinetics in these samples are the shortest, indicating a higher probability of the charge transfer process compared to other films. Consequently, the main factor affecting the transfer efficiency is the ligand length, which determines the distance between neighboring QDs, rather than their chemical structure or uniformity of QDs distribution within the sample.

In order to analyze evolution of characteristic times of luminescence decay along with transition from a solution to QD-film and from longer to shorter ligands in QD-films, we describe kinetics of charge carriers and excitons by the following set of equations:

$$\frac{dP}{dt} = E/\tau_d - k_m P^2 \tag{2}$$

$$\frac{dE}{dt} = -E\left(\frac{1}{\tau_d} + \frac{1}{\tau_0}\right) + k_m P^2, \tag{3}$$

where E and P are concentrations of excitons (which is proportional to luminescence intensity) and charge carriers (holes and electrons), respectively; τ_0 is a recombination time in solution (i.e. lifetime of excitons in isolated QDs), and τ_d is a transfer time of a carrier to the neighbor QD. Eq. 2 assumes that charge carriers appear due to decay of excitations, and disappear due to formation of excitons after the meeting with opposite charge, see the second term with the rate constant k_m (generation is supposed to be spatially uniform, hence transport term is absent). Eq. (3) means that decay of excitons results from production of charge pairs and recombination, the rate of the latter is the same as in solution. Biexponential character of the luminescence decay, see Eq. (1), provides that E is a solution of the differential equation $\ddot{E} + 2\beta\dot{E} + \omega^2 E = 0$, $\beta^2 - \omega^2 \geq 0$, where β and ω are some constants. Comparing this equation and Eq. (3) after differentiation in time, one require $\omega^2 E \approx -2k_m P\dot{P}$. Analysis of the system (2), (3) under this approximation yields

$$\frac{1}{\tau_1} + \frac{1}{\tau_2} = \frac{1}{\tau_d} + \frac{1}{\tau_0} \tag{4}$$

Substitution of measured values of τ_0, τ_1 , τ_2 for the case of ligands OA leads to the conclusion, that $\tau_d \approx \tau_1$ (more precisely, $\tau_d \leq \tau_1$). Eq. (4) is consistent with experimental result $\tau_0 > \tau_2$.

Since $\tau_1 \approx \tau_2$ for the case of ligands HDT and hydrazine, and $\tau_0 \gg \tau_d$, eq. (4) yields

$$\tau_d \approx \tau_1/2 \tag{5}$$

for these ligands. Providing hopping nature of charge separation, we assume

$$\tau_d = t_0 \exp(2\gamma l), \tag{6}$$

where γ is an inverse localization radius, l is the length of a ligand molecule, and t_0 is an attempt-to jump time. Substitution of τ_d values from Eqs (4), (5) to Eq. (6), and $l \approx 2$, 1 and 0.5 nm for OA, HDT and hydrazine, respectively, yields ${\gamma_{opt}}^{-1} \approx 1.2$ nm. One has to note, that the estimation of the same value from steady-state conductivity measurements [16] results in considerably smaller value, ${\gamma_{cond}}^{-1} \approx 0.4$ nm. , which is more consistent with the hopping nature of charge transfer. One has remind, however, that the measurements of luminescence decay shows the transient process in a disordered media on a short time scale. Charge transport should be involved in this process in an extremely non-equilibrium (dispersive) mode, while conductivity measurements [16] deals with the transport in quasi-equilibrium mode. In the first case, an average hopping frequency should be much larger, because the fastest jumps realized on a short time scale after the generation pulse. That means that the 1) hopping distances are considerably scattered (positional disorder), 2) the shortest distances are involved in hopping on a short time scale, and 3) thermally activated (hence delayed) hopping is a rate-limiting step in a quasi-equilibrium mode, due to energetic disorder. Thus, the $\boldsymbol{{\gamma_{opt}}^{-1} \approx 1.2}$ nm seems to be an apparent value, rather the shorter lengths l should be considered.

5. Conclusion

Using PbS QD solids, we demonstrated that the shorter the length of ligand molecules the faster the rate of luminescence decay. This length is considered as a measure of the mean inter-QD distance in QD solid. We propose that the reason of the correlation is the increase of an additional channel of exciton's decay, namely the exciton dissociation by the means of hopping of a charge carrier to the neighbor QD. This result agrees well with the recently reported conclusion that photocurrent increases exponentially with decreasing ligand length [16].

6. Acknowledgements

The work was supported by "improving the competitiveness" program of the National Research Nuclear University «MEPhI». Preparation of QD solids was supported by the Russian Foundation for Basic Research, grant no. 16-29-09623 (OFI_M).

7. References

[1] Choi J H, Fafarman A T, Oh S J et al 2012 *Nano Lett.* **12** 2631
[2] Chistyakov A A, Zvaigzne M A, Nikitenko V R et al 2017 *J. Phys. Chem. Lett.* **8**
[3] Pathak D, Wagner T, Adhikari T, Nunzi J M 2015 *Synth. Met.* **200** 102
[4] Kramer I J, Sargent E H 2014 *Chem. Rev.* **114** 863
[5] Dayneko S, Lypenko D, Linkov P et al 2016 *Materials Today: Proceedings* **3** 211
[6] Dayneko S, Lypenko D, Linkov P et al 2014 *Proc. SPIE* **9270** 927009
[7] Guyot-Sionnest P 2012 *J. Phys. Chem. Lett.* **3** 1169
[8] Gilmore R H, Lee E M Y, Weidman M C, Willard A P, Tisdale W A 2017 *Nano Letters* **17** 893
[9] Dayneko S, Linkov P, Martynov I et al 2016 *Physica-E: Low-dimensional Systems and Nanostructures* **79** 206
[10] Dayneko S, Tameev A, Tedoradze M et al 2013 *Appl. Phys. Lett.* **103** 063302
[11] Yang J, Wise F W 2015 *J. Phys. Chem. C* **119** 3338
[12] Xu F, Gerlein L, Ma X, Haughn C, Doty M and Cloutier S 2015 *Materials* **8** 1858
[13] Choi J J, Luria J, Hyun B R et al 2010 *Nano Letters* **10** 1805
[14] Luther J M, Law M, Qing Song, Craig L et al 2008 *ACS Nano* **2** 271
[15] Zarghami M H, Liu Y, Gibbs M et al 2010 *ACS Nano* **4** 2475
[16] Zvaigzne M A, Alexandrov A E, Samokhvalov P S et al 2017 *Tech. Phys. Lett.* **44**

IOP Conf. Series: Materials Science and Engineering **475** (2019) 012029 doi:10.1088/1757-899X/475/1/012029

Quantum Hall transitions in the presence of Landau levels mixing in n- InGaAs/InAlAs structures

S V Gudina[1], Yu G Arapov[1], E V Ilchenko[1], V N Neverov[1], A P Savelyev[1], S M Podgornykh[1], N G Shelushinina[1], M V Yakunin[1], I S Vasil'evskii[2] and A N Vinichenko[2]

[1]M.N. Miheev Institute of Metal Physics of Ural Branch of Russian Academy of Sciences, S. Kovalevskaya str. 18, 620990, Yekaterinburg, Russia
[2]National Research Nuclear University MEPhI, Kashirskoe shosse, 31, 115409, Moscow, Russia

Corresponding author's e-mail address: svpopova@imp.uran.ru

Abstract. We provide a systematic measurement of the longitudinal ρ_{xx} and Hall ρ_{xy} resistivities of quantum Hall transition in a two-dimensional electron system $In_{0.9}Ga_{0.1}As/In_{0.81}Al_{0.19}As$ with strong spin-orbit coupling. For half-integer filling factors the linear temperature dependence of the effective quantum Hall effect plateau-to-plateau transition width $\Delta B(T)$ is observed in contrast to scaling behavior for systems with short-range disorder. The recent prediction that the width of transition region remains finite when extrapolated to zero temperature, resulting from Landau level mixing, is more preferable. The shift of the transition point in magnetic field with the temperature is found to originate from the mixing between Landau levels due to the inelastic scattering.

1. Introduction

The critical behavior near extended states in the quantum Hall effect is a subject of great interest [1-50]. The integer quantum Hall effect (QHE) demonstrates universal scaling behavior [1-4] in accordance with a single delocalized quantum critical point (E_c) [5-12] between two adjacent plateau states. Near the quantum critical point, the localization length ξ diverges as a power law $\xi(E) \sim |E_F - E_c|^{-\gamma}$ [6, 7], with E_F being the Fermi energy and γ the localization length exponent. Physical quantities such as conductance follow the one-parameter scaling law as a function of the scaling variable $(L/\xi)^{1/\gamma} \sim L^{1/\gamma}|E_F - E_c|$, L being the system length. However, for experiments done at finite temperature, the effective system length for coherent transport is determined by a finite length scale $L_\varphi \sim T^{-p/2}$ representing the dephasing effect [1, 7] due to inelastic scattering. T is the temperature and the parameter p is known as the inelastic-scattering exponent. Thus the scaling parameter for finite temperature systems appears to be $T^{-\kappa}\Delta E$ with $\kappa = p/2\gamma$ [1-7].

It has experimentally been found that the half width ΔB of the ρ_{xx} peak between adjacent Hall plateaus and the slope $d\rho_{xy}/dB$ of the Hall resistance ρ_{xy} show a power-low behavior $\Delta B \sim T^{\kappa}$ and $d\rho_{xy}/dB \sim T^{\kappa}$.

While some experimental works starting with [2] find κ as universal and equal ≈0:42, independent of material and Landau level number, in other reports the universality of κ is questioned [13-26].

The crossing of spin-split Landau levels with the same number and different spin index leads to interesting physics. When Landau levels are mixing, the interaction of electrons in different Landau levels complicate this problem and draw into question whether the above universality of scaling behavior is unchanged [23-26]. On the one hand, based on different models and calculation approaches, many theoretical works [23, 24] reach the conclusion that Landau level mixing will not change the universality of the quantum Hall phase transition. However, Xiong *et al.* give a striking and fundamentally different prediction that a narrow band of extended states is formed near each Landau level center due to interband mixing and leads to nonscaling behaviors [25, 26]. On the other hand, up to now the focus has been mainly theoretical with a lack of systematic studies on the experimental side. There are only a few experiments on one special case of Landau level mixing: the Landau levels associated with opposite spin orientations are not resolved [29-31]. Thus, a unified picture is still lacking for quantum Hall phase transition in the presence of Landau level mixing. In contrast to earlier experiments, here we measure the transition directly in the spin resolved Landau level mixing regime.

2. Experimental details

Semiconductor heterostructures with an InAlAs/InGaAs/InAlAs quantum well and metamorphic buffer of varied composition $In_xAl_{1-x}As$ were grown by molecular-beam epitaxy using a RIBER Compact 21 system at the Institute of Functional Nuclear Electronics of the National Research Nuclear University "MEPhI". InP substrates (Wafer Technology Co.) with (100) orientation were used. The layer sequence and parameters of the sample are given in the [49]. Electron density n=6.8·10^{15} m^{-2}, mobility μ = 20 m^2/V·s.

The composition profile (indium concentration x) in the $In_xAl_{1-x}As$ metamorphic buffer was technologically specified as linear; a superlattice was introduced into the heterostructure to suppress dislocation growth into the active region. The metamorphic buffer was completed with an inversion step with linearly decreasing x to prevent elastic-strain penetration into the active region. Then a thick $In_xAl_{1-x}As$ pseudo-substrate layer was grown as the substrate for subsequent active layers of the $In_yGa_{1-y}As$ quantum well (QW), an $In_xAl_{1-x}As$ spacer layer, a δ-layer of dopant Si, an $In_xAl_{1-x}As$ barrier layer, and an undoped $In_yGa_{1-y}As$ protective layer. The sample on the InP substrates contained a pseudomorphically strained QW whose lattice parameter exceeded that of the barrier layers. Silicon was used as the donor impurity and the Si-atom concentration was 1.5·10^{12} cm^{-2}.

The sample is made in the form of a double Hall bar. Longitudinal ρ_{xx} and Hall ρ_{xy} magnetoresistances were measured simultaneously in magnetic fields up to 9 T and temperatures from 1.8 K to 30 K using Quantum Design standard set up and in magnetic fields up to 12 T and temperatures from 0.4 K to 1.6 K using ^{3}He Oxford Instruments system. Measurements were carried out on the equipment of the Collaborative Access Center "Testing Center of Nanotechnology and Advanced Materials" of the Miheev Institute of Metal Physics of the Ural Branch of the Russian Academy of Sciences.

3. Magnetotransport results and discussion

3.1. Longitudinal and Hall resistances at magnetic fields

Figure 1 shows the magnetotransport measurements of our sample under perpendicular magnetic fields up to B=9.0 T at T=1.8-10.0 K. A well-developed Landau quantization at the integer quantum Hall regime is observed: well pronounced Shubnikov-de Haas oscillations in ρ_{xx} and quantized plateaus in ρ_{xy}. In the magnetic structure, sharp quantum Hall steps are clearly seen for both even and odd filling factors up to $\nu\approx 11$ (at T=1.8 K, fig.1). At the same time, in the case of the ordinary heterostructures, odd plateaus are resolved only up to ν = 3. This indicates enhanced spin splitting of Landau levels in the sample due to the spin-orbit interactions.

In Figure 2 we represent a half-width of ρ_{xx} peaks defined as the interval ΔB in B between the extreme in dρ_{xx}/dB and the point where dρ_{xx}/dB is equal to zero. The experimental information was analyzed from the low- and the high-field sides of transition point. These two types of a half-width of ρ_{xx} peaks correspond to originally different gaps in LL spectrum. ΔB is a measure for the fraction of Landau-level states, which contribute to the conduction at finite T. It can be seen that experimental data for 3-4 transition are in a good agreement for both conditions $B>B_c$ and $B<B_c$. Measured data for 4-5 and 5-6 transitions are slightly diverges when temperature increases. However, the slope remains the same. It is clearly seen that for odd numbers of QHE plateaus the scaling regime completes earlier than for even ones. Calculated values of critical exponent are in complete accord with the experimental findings obtained in [49], where we measured the half-width ΔB of R_{xx} curve as a function of temperature.

We find that ΔB vanishes according to both the power law T^κ in the temperature interval from ~3 K to 10 K and the linear dependence $\alpha T + \beta$ at T< (6÷8) K. The κ values are, within the estimated confidence interval, distinctly different from that typically associated with an Anderson-type transition, namely, κ = 0.42 [2-4], but they correlate well with experimental results for systems with large-scale impurity potentials [3, 4, 15-19].
Figure 3 shows the evolution of B_C with T varying. B_C was determined as the position of dρ_{xy}/dB maximum for appropriate transition at fixed T. It can be seen that B_C tends to a fixed value at low temperature limit.

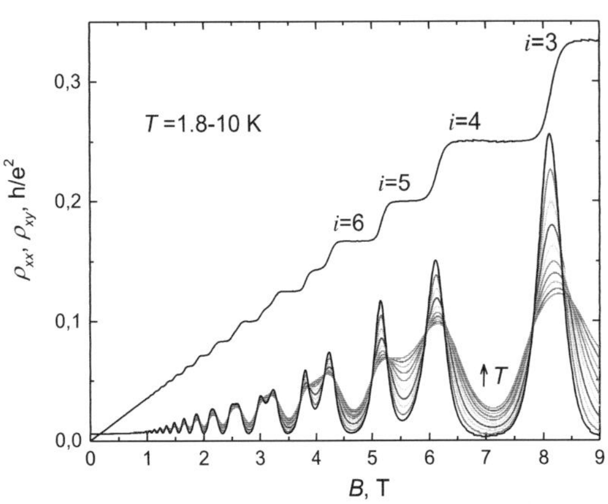

Figure 1. Magnetic field dependences of ρ_{xx} (T=1.8-10 K) at fixed temperatures (arrow shows the growth of T) and ρ_{xy} (T=1.8 K).

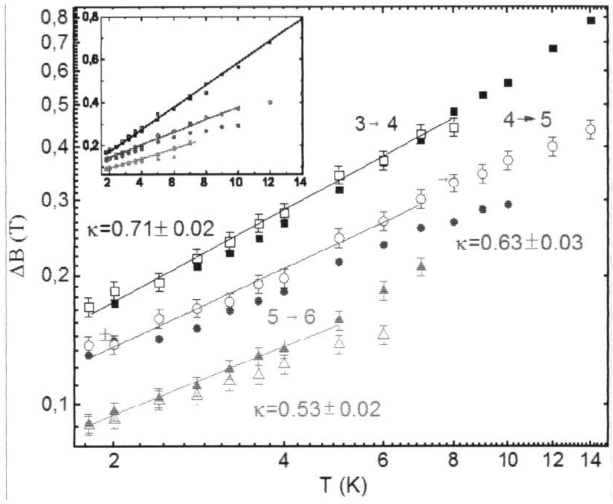

Figure 2. Temperature dependences of the half-width ρ_{xx} peaks for transitions between plateaus 3→4, 4→5 and 5→6, filled symbols – $B< B_c$, open symbols – $B>B_c$. The solid lines represent the $\Delta B \propto T^{\kappa}$. Inset shows ΔB(T) in linear scale, solid lines – $\Delta B \propto \beta + \alpha T$.

3.2. The width of the extended state band

According to the standard scaling theory [1], the Hall ρ_{xy} and the longitudinal ρ_{xx} resistivities will have power-law behaviors with temperature: the former has a slope at the critical point as $d\rho_{xy}/dB\,|_{Bc}\sim T^{-\kappa}$, while the latter has a half-width of the transition region as $\Delta B \sim T^{\kappa}$. In terms of this conception the transition regions between adjacent QHE plateau and the width of appropriate $\rho_{xx}(B)$ peaks should become narrower and narrower as the temperature approaches zero.

The value of γ in the context of short-range potential and without regard to the e-e interaction was theoretically obtained to be 2.3 (see, for example, review [7, 46]). Afterwards, the generalized theory that takes into account the e-e interaction as well as spin-polarized and spin-degenerate electron states was constructed. In fact, this theory introduces special scaling parameter considering the renormalization of interaction constant and subsequently, new values of the critical exponents γ and p were estimated to be 2.75 and 1.35±0.15 respectively.

There were the several bright experimental studies where the universal value of exponent κ=0.42 was established for heterostructures with short-range impurity potential [2-4, 15, 16, 27]. On the other hand, the universality of the exponent κ was questioned in other series of experiments [13-21].

To clarify the influence of the interaction and spin effects on the universality of critical exponents values the idea to study the half-width of resistance (conductivity) peaks on the different (high-field and low-field) sides from the critical point was arose [22, 50]. The purpose was to separate the effects which are better seen and originated from the Zeeman and cyclotron gaps.

Meisels *et al* analyzed the longitudinal conductivity σ_{xx} peak between adjacent IQHE states with respect to temperature and frequency scaling [50]. The critical behavior of σ_{xx} was studied on the different sides of the peak width. The authors separated non-universal impurity-induced (κ > 0.4) effects from universal behavior (κ ≈ 0.4) within the localization model of the QHE. It is reported that for ν < 1.5 the presence of attractive impurities giving rise to at least a shoulder in the density of states (DOS) and this asymmetry in the DOS leads to a strong increase of the value of κ (κ > 1) on the high-field side of the σ_{xx} peak. All the higher-ν direct current peaks showed a tendency toward κ ≈ 0.2, when the spin splitting was not resolved.

Some interesting effects were highlighted for peaks belonged to different spin direction [50]. For example, the temperature dependence of the spin-up peak widths was enhanced (κ≈ 0.3) due to the overlap with the spin-down peaks which increases with temperature since the spin-split subbands become more equally occupied (reduced g-factor enhancement). Also such a growth of the spin-up peaks is probably due to the effect of spin-orbit scattering [47]. For the peaks ν≈ 7/2 (low-field), 11/2 (low-field), 5, 7, 9, the subbands were more equally occupied (effective value of g-factor is closer to its bare value), then the temperature dependence was reduced (κ≈ 0.2).

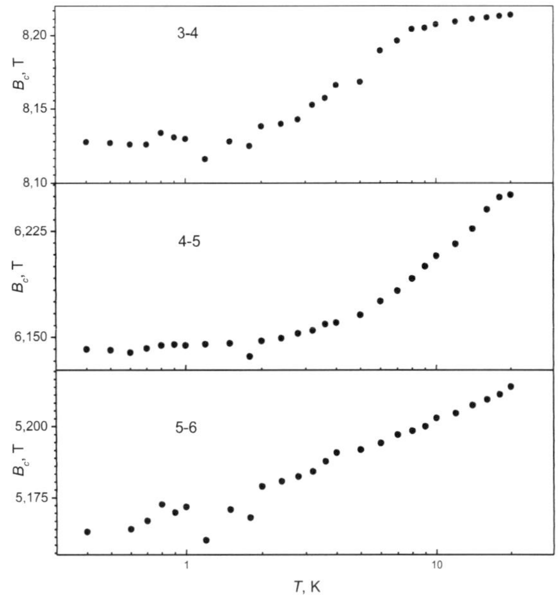

Figure 3. The temperature dependence of critical magnetic field B_C of quantum phase transitions (log-linear plot) for the sample under study (with field variation ΔB: for the transition 3-4 ΔB =90 mT, for the 4-5 – ΔB=108mT, for the 5-6 – ΔB =43 mT).

Theoretically Xiong *et al.* claim that the strong inter-Landau band mixing can result a finite bandwidth which is compared to the spacing of two adjacent Landau bands [25, 26]. They proved convincingly that a narrow metallic region is formed near each Landau level center due to Landau level mixing [25]. The theory was developed further to include interband overlap [26]. They show that for clean samples the extended state in each Landau level remains a single point at zero temperature. In contrast, for dirty samples the extended state in each Landau level can form a narrow band instead of a single point.

In the work of Li Wang *et al* [22] the Xiong's theory [25, 26] was experimentally checked using exploration of the width of quantum Hall transition region in the presence of spin resolved LL mixing regime. The authors investigated a 24-nm wide single GaAs quantum well bounded on each side by Si δ-doped layers of AlGaAs. They discussed the half-width ΔB of R_{xx} curve as a function of temperature and additionally the data was analyzed from low- and high-field sides of transition region. The value of κ was calculated to be 0.72. The critical exponents in the left- and right-hand sides of transition region was extracted as $\kappa_L = 0.72$ and $\kappa_R = 0.75$. The experimental results demonstrated that the standard scaling behavior remains unchanged even with strong Landau level degeneracy. The authors concluded that their experiment may correspond to the case of clean samples, in Xiong's terms, where the extended state in each Landau level remained a single point at zero temperature (mobility is high as the order of 10^5 cm^2/V·s).

Let us discuss our results and compare them with the experimental findings of [22, 50]. It was mentioned in previous section that figured out values of critical exponent are in a good agreement with the experimental results obtained in [49], similar to Wang's research, where estimated values of κ are equal for two approaches within the experimental uncertainty. In the present work, the values of κ are the same within the measurement accuracy for different types of gaps, in contrast to Meisels's work [50].

In [17, 18] a transport regime distinct from the critical scaling behavior was reported to exist asymptotically close to the transition at very low temperatures. The effective transition width of GaAs/AlGaAs and InGaAs/InP samples appears to vary as $\alpha T + \beta$ rather than exhibiting T^{κ} scaling behaviour. This means that at $T = 0$ the transition has a finite width.

To estimate the width of a band of delocalized states in our n-$In_{0.9}Ga_{0.1}As/In_{0.81}Al_{0.19}As$ samples we have analyzed magnetoresistance data in transition region between the filling factors 3-4, 4-5, 5-6 QHE plateau. In figures 2 (inset) the $\Delta B(T)$ dependences on a linear scale for the investigated sample are presented. It is seen that the data can be satisfactorily described by a power law $\Delta B \sim T^{\kappa}$. On the other hand, the data are far more compatible with a linear dependence $\Delta B(T) = \alpha T + \beta$ with $\beta = 0.16$ T, $\beta/\alpha = 1.65$ K for 3-4 transition, $\beta = 0.17$ T, $\beta/\alpha = 3.1$ K for 4-5 transition and $\beta = 0.12$ T, $\beta/\alpha = 3.3$ K for 5-6 transition.

As pointed out the ratio β/α defines a temperature T^* that is found to be characteristic of the material system. T^* turned out to be close to 0.5 K for InGaAs/InP samples and 50 mK for GaAs/AlGaAs samples [17, 18]. For Ge/GeSi samples studied in [21] the characteristic temperature is about 2.5 K (2.3–2.8 K).

What a reason for the linear $\Delta B(T) = \alpha T + \beta$ dependence? The most natural reason for the linear $\Delta B(T)$ dependence, namely, the thermal broadening of a quantum critical phase transition, is suggested and confirmed by calculation in [19]. It is shown there that the thermal broadening Fermi-Dirac distribution function $f(E)$ gives the linear increase of $\Delta B(T)$.

The authors [21] think that the answer to the main question about the finite $T \to 0$ width of QHE transitions may be found in the works treating the influence of Coulomb interactions on the screening of smooth disorder potential [33, 34]. The theory includes screening within the Thomas –Fermi approximation appropriate for smooth disorder.

When Landau levels are mixing, the interaction of electrons in different Landau levels with the same number and oppositely directed spins complicate this problem, especially in the case of nanostructures with strong spin-orbit interactions [25, 26, 38, 39, 47]. Particularly the half-width of the transition region can have a linear dependence with temperature $\Delta B = \alpha T + \beta$, which remains finite even when the temperature T approaching zero.

For comparing both theoretical predictions, we have plotted two different fits in Fig. 2. It is obvious that in the measurable regime the power law and linear fits are both good, where the fitting parameters are $\kappa = (0.60\text{-}0.76)$ and $\beta = (0.12\text{-}0.16)$T in the two methods, respectively. The deviation between the two fits lies within the experimental uncertainty. In other words, it is difficult to find the obvious evidence to support the two separated critical points prediction even in the presence of strong Landau level mixing.

Thus, our experiment may correspond to the presence of a band of delocalized states, as a consequence of strong mixing and overlapping Landau levels with the same number and opposite spin directions.

3.3. The temperature dependence of critical magnetic field B_C of quantum phase transitions

The behaviors of critical magnetic field B_C for all the three transitions are summarized in Fig.3, which provides a possible verification of the scaling theory. In both the $3 \to 4$ and $4 \to 5$ plateau-plateau transitions, the critical fields B_C rapidly moves from the low temperature value 1.0-1.8K with field variation $\Delta B_c = (43.0 \div 108.0)$ mT at 20.0 K for different transitions. However, for $5 \to 6$ transition, the critical fields B_C is much smoother when increasing the temperature, with the field variation as low as $B = 43$ mT at 20.0 K.
Such a variation of B_C can be expected from the irrelevant finite size correction at high temperature [36]. Also, the absence of the independent temperature percolation point ($B_c(T)$) may be due to scattering events [37]. The magnetic-field shift $B_C(T)$ (or energy shift $E_c(T)$) of the saddle point for increasing temperature can be explained by the increase of the scattering between electrons moving along percolation lines belonging to the highest Landau level (N=1) or the N=2 electrons rotating along closed loops around the potential hills. This shift due to Landau-level mixing of delocalized states, named ''weak levitation,'' has been predicted recently [38, 39].

The authors [38, 39] employed a two-channel version of Chalker-Coddington network model [6], which was previously used to study a spin-split Landau level. Electrons move along unidirectional links that form closed loops in analogy with semiclassical motion on contours of constant potential. Nodes correspond to regions in space where two classical contours approach one another, i.e., nodes are saddle points in the potential. At nodes, tunneling must complement the semiclassical motion so that scattering at nodes couples states on neighboring links. The assumption that each link carries current only in one direction implies that the wave packets are sufficiently localized in the transverse direction, i.e., the magnetic length is small in comparison with the spacing of nodes or with the correlation length of the potential fluctuations. The network model is therefore a strip whose width has half-integer number of links with two channels per link and scattering of states on two neighboring links is allowed at quarter part of nodes.

In [38, 39] was shown how the the energy of the delocalized state are gone upward from the position $h\omega_c(n + 1/2)$ due to Landau level mixing, ω_c is the cyclotron frequency, n=0, 1, 2.... This shift is relatively small in a smooth potential even when the peaks in the density of states corresponding to different Landau level are not well resolved.

Up to now it is considered a single-particle picture of delocalization. The authors completely neglected the effects of screening caused by electron-electron interactions. At the same time, in a smooth potential the electron-electron interactions drastically affect the distribution of electrons within the plane. In fact, when the potential is smooth enough, the equipotentials are separated by incompressible strips [40, 41]. In this situation, one should speak not about the energy position of delocalized state but rather about the critical filling factor at which delocalization occurs. It was argued in [37] that by replacing equipotentials with edge excitations, the many-electron problem can be effectively reduced to the Chalker-Coddington model [6, 7]. In the framework of this scenario the theory [42] predicts that the critical filling factor is larger than $n + 1/2$ due to Landau level mixing.

We attribute the behaviors of critical magnetic field B_C for all the three transitions to the large spin-orbital interaction effect in the n- InGaAs/InAlAs nanostructures, which has been known in this system [43-45]. The Landau levels are spin non-degenerate due to the Zeeman splitting, and the Landau levels with filling factor $2n - 1$ and $2n$ are also strongly coupled due to the spin-orbital coupling effect. Because the Landau energy is larger than the Zeeman splitting energy, the inter-level coupling between the third and fourth plateau is more effective than the other two transitions, which causes a shorter localization length ξ, weakens the irrelevant finite size correction [5-7], and leads to a higher precision of the critical B_C. Moreover, although in the theoretical aspect the spin-orbital coupling effect can lead to delocalized states, the large magnetic field change the system to the unitary class [46], which only contains delocalized states at the center of Landau levels.

As mentioned above the specific peculiarity of studied structures is the presence of strong spin-orbit interaction, which affects the spin splitting of Landau levels. In [47, 48] was considered the case where the electronic system with a smooth potential disorder and strong spin-orbit interaction in the two mode conditions widened Zeeman Landau level much overlap. It turned out that the spin-orbit interaction causes changes in the percolation grid, which leads to an increase of the dissipative conductivity at finite T, when the Fermi level lies between the energies of two delocalized states at the centers of the widened Zeeman Landau levels.

4. Conclusion

In conclusion, we have measured the temperature dependences of the longitudinal ρ_{xx} and Hall ρ_{xy} resistivities in the region of quantum Hall transition. We focus on the two Zeeman Landau levels mixing case. For half-integer filling factors the linear temperature dependence of the effective quantum Hall effect plateau-to-plateau transition width $\Delta B(T)$ is observed in contrast to scaling behavior for systems with short-range disorder. Our experiment may correspond to the presence of a band of delocalized states, as a consequence of strong mixing and overlapping Landau levels with the same number and opposite spin directions.

Acknowledgments

The work was carried out using equipment of the Center for Collective Use of the National Research Nuclear University MEPhI "Heterostructural Microwave Electronics and Wide-Gap Semiconductors". Experiments were carried out at the Collaborative Access Center "Testing Center of Nanotechnology and Advanced Materials" of the M.N.Miheev Institute of Metal Physics of the Ural Branch of the Russian Academy of Sciences. The research was carried out within the state assignment of Russian Ministry of Science and High education (theme "Electron" No. AAAA-A18-118020190098-5) and Complex Program of RAS Ural Branch 18-10-2-6, supported in part by RFBR (projects Nos 18-02-00172, 18-02-00192).

References

[1] Pruisken A M M 1988 *Phys.Rev. Lett.* **61** 1297
[2] Wei H P, Tsui D C, Paalanen M A and Pruisken A M M 1988 *Phys. Rev. Lett.* **61** 1294
[3] Li W, Xia J S, Vicente C, Sullivan N S et al 2010 *Phys. Rev. B* **81** 033305
[4] Li W, Vicente C L, Xia J S, Pan W, Tsui D C et al 2009 *Phys. Rev. Lett.* **102** 216801
[5] Huckestein B and Kramer B 1990 *Phys. Rev. Lett.* **64** 1437
[6] Chalker J T and Coddington P D 1988 *J. Phys. C* **21** 2665
[7] Huckestein B 1995 *Rev. Mod. Phys.* **67** 357
[8] Kivelson S A, Lee D-H and Zhang S-C 1992 *Phys. Rev. B* **46** 2223
[9] Huo Y, Hetzel R E and Bhatt R N 1993 *Phys. Rev. Lett.* **70** 481
[10] Wang Z Q, Jovanovic B and Lee D-H 1996 *Phys. Rev. Lett.* **77** 4426
[11] Yang K and Bhatt R N 1996 *Phys. Rev. Lett.* **76** 1316
[12] Yang K and Bhatt R N 1999 *Phys. Rev. B* **59** 8144
[13] Balaban N Q, Meirav U and Bar-Joseph I 1998 *Phys. Rev. Lett.* **81** 4967

[14] Wakabayashi J, Yamane M and Kawaji S 1989 *J. Phys. Soc. Japan.* **58** 1903
[15] Koch S, Haug R J, K v Klitzing and Ploog K 1991 *Phys. Rev. Lett.* **67** 883
[16] Koch S, Haug R J, K v Klitzing and Ploog K 1991 *Phys. Rev. B* **43** 6828
[17] Shahar D, Hilke M, Li C C, Tsui D C, Sondhi S L et al 1998 *Sol. Stat. Comm.* **107** 19
[18] Shahar D, Tsui D C, Shayegan M, Shimshoni E and Sondhi S L 1996 *Phys. Rev. Lett.* **79** 479
[19] Coleridge P T 1999 *Sol. Stat. Comm.* **112** 241
[20] Schaijk R T F van, Visser A de, Olsthoorn S et al 2000 *Phys. Rev. Lett.* **84** 1567
[21] Arapov Yu G, Alshanskii G A, Harus G I, Neverov V N et al 2002 *Nanotechnology* **13** 86
[22] Wang L, Tu T, Zhou C, Zhao Y-J, Guo G-C, Guo G-P 2013 *Modern Phys. Lett.* **27** 1350202
[23] Wang Z, Lee D-H and Wen X-G 1994 *Phys. Rev. Lett.* **72** 2454
[24] Lee D K and Chalker J T 1994 *Phys. Rev. Lett.* **72** 1510
[25] Xiong G, Wang S-D, Niu Q, Wang Y, Xie X C et al 2006 *J. Phys.: Condens. Matter* **18** 2029
[26] Xiong G, Wang S-D, Niu Q, Wang Y and Wang X R 2008 *EPL* **82** 47008
[27] Arapov Yu G, Gudina S V, Klepikova A S, Neverov V N et al 2013 *JETP* **117** 144
[28] Gudina S V, Arapov Yu G, Saveliev A P, Neverov V Net al 2016 *Semiconductors* **50** 1641
[29] Hwang S W, Wei H P, Engel L W, Tsui D C and Pruisken A M M 1993 *Phys.Rev. B* **48** 11416
[30] Engel L W, Shahar D, Kurdak C and Tsui D C 1993 *Phys. Rev. Lett.* **71** 2638
[31] Wei H P, Engel L W and Tsui D C *Phys. Rev. B* 1994 **50** 14609
[32] Pruisken A M M, Scoric C B and Baranov M A 1999 *Phys. Rev.B* **60** 16838
[33] Efros A L 1992 *Phys. Rev. B* **45** 11354
[34] Cooper N R and Chalker J T 1994 *Phys. Rev. B* **48** 4530
[35] Lee D K and Chalker J T 1994 *Phys. Rev. Lett.* **72** 1510
[36] Wang X *et al.* 2016 *Phys. Rev B* **93** 075307
[37] Gusev G M *et al.* 1998 *Phys. Rev B* **58** 4638
[38] Shahbazyan T V and Raikh M E 1995 *Phys. Rev. Lett.* **75** 304
[39] Kagalovsky V, Horovitz B and Avishai Y 1995 *Phys. Rev. B* **52** R17044
[40] Efros A L 1988 *Solid State Commun.* **67** 1019
[41] Chklovskii D B, Shklovskii B I and Glazman L I 1992 *Phys. Rev. B* **46** 4026
[42] Lee D-H, Wang Z Q and Kivelson S 1993 *Phys. Rev. Lett.* **70** 4130
[43] Usher A, Nicholas R J, Harris J J and Foxon C T 1990 *Phys. Rev.B* **41** 1129
[44] Leadley D R, Nicholas R J, Harris J J and Foxon C T 1998 *Phys.Rev. B* **58** 13036
[45] Maude D K, Potemski M, Portal J C, Henini M et al 1996 *Phys. Rev. Lett.* **77** 4604
[46] Evers F and Mirlin A D 2008 *Rev. Mod. Phys.* **80** 1355
[47] Polyakov D G, Raikh M E 1995 *Phys.Rev.Lett.* **75** 1368
[48] Avishai Y, Meir Y 2002 *Phys.Rev.Lett.* **89** 076602
[49] Gudina S V, Arapov Yu G, Ilchenko E I, Neverov V N et al 2018 *Semiconductors* **52** 1551
[50] Meisels R, Kuchar F, Belitsch W, and Kramer B 1999 *Microelectronic Engineering* **47** 23

Etching technology in formation of self-organized aluminum mesh for sensor and display applications

V Zhylinski[1], N Bagamazava[1], V Bezborodov[1], A Smirnov[2], A Chernik[1] and I Zharski[1]

[1]Belarusian State Technological University, Minsk, Sverdlova Str., 13a, 220006, Belarus
[2]Belarusian State University of Informatics and Radioelectronics, P. Brovki Str. 6, 220013, Belarus

Corresponding author's e-mail address: zhilinski@yandex.ru

Abstract. The paper describes the investigations the temperature and concentration influence on etching of nanoporous alumina in technology of production of self-organized aluminum mesh for sensor and display applications. The pyramidal aluminum meshes was obtained by anodizing of thin film sputtered on sodium glass in oxalic acid solution at voltage of 50 V and then chemical etching of porous alumina in the mixture of phosphoric and chromic acid. It has been shown that the pyramidal morphology of Al meshes can be prepared by chemical etching at temperature of 60°C. This pyramidal morphology of Al mesh can provide good alignment properties and can be one of the promising solutions for production of sensory microsystems and displays without indium-tin oxide films.

1. Introduction

It has been reported that the electrochemical anodization of aluminum on a glass substrate can provide Al meshes with high optical transparency, a low electrical resistivity and good alignment properties for liquid crystal (LC) molecules on the substrate surface. This can be one of the promising solutions for the design and production of next generations of displays and sensors without indium-tin oxides (ITO) [1-4]. The anodization of aluminum is accompanied by the formation of porous alumina layer, which is necessary to remove for production of Al mesh [5].

In the last works [5, 6] for the formation of nanoporous alumina layers on glass substrate, we used the electrochemical anodization of aluminum layers at voltage of 50 V in oxalic acid solution. We founded optimal conditions of the anodization in oxalic acid solution, which allowed preparing nanostructured surfaces with a good optical transparency and conductivity, and good alignment properties of LC molecules on the substrate surface [5-8]. It should be noted that the short time of the anodization did not allow producing good quality transparent nanoporous alumina layers and long-continued time could lead to the formation of the layers with a high electrical resistivity and a low conductivity. Using trade-off between the transparency and conductance of the layers we obtained resistance of 10-20 Ω per square sheet for holes (sphere) with radius of 60-70 nm and interholes distance of 100 nm [7]. Moreover, the transmittance of such Al meshes depends on conditions of etching of anodic aluminum oxide. The research of etching process of anodic aluminum oxide is practically important problem [1-4] of development of technology of formation of self-organized aluminum mesh for sensor and display applications.

In this paper, we report about investigation of the temperature and concentration influence on etching of nanoporous alumina, which has been obtained by anodization in oxalic acid, for the technology of formation of transparent Al meshes.

2. Materials and Methods

Thin Al layers (thickness of 1000 nm) was sputtered on sodium glass by using magnetron spraying. The specimens with dimensions of 1 × 1.5 cm were cut off from sodium glass with thin Al layers and then they were degreased in ethanol. The area of 1.0 cm^2 was selected by insulation of the rest of the sample. An acid resistant paint was used for insulation [5].

Nanoporous alumina layers were made on the surface of aluminum by a simple anodization, which was carried out in 0.3 M oxalic acid at voltage of 50 V and temperature of 10°C [6]. We used DC power supply brand MNIPI B5-78/7 for anodization. The current drop determinate the duration of anodization. The acid resistant paint was dissolved in acetone after anodization [6].

Oxide films were removed by chemical etching in a mixture of H_3PO_4 and H_2CrO_4. Experiments were performed at different temperatures (40, 60 and 80 °C) and different concentration of H_3PO_4 (0.3, 0.6 M) and H_2CrO_4 (0.2 and 0.4 M) in order to investigate the temperature effect on the conductivity of transparent Al meshes.

The optical transmittance of Al meshes on glass was measured by digital spectrophotometer Metash UV-6000pc in the wavelength range from 300 to 1200 nm. The morphology and cross-sections of the Al meshes were examined by Hitachi S-806 and Zeiss EVO MA 15 scanning electron microscopes.

3. Results and Discussion

For the investigation of etching in mixture of phosphoric and chromic acid were estimated the mass losing under dissolution of layer of the nanoporous alumina (Figure 1) which had been formed by anodization in 0.3M oxalic acid at voltage of 50 V and temperature of 10°C [6].

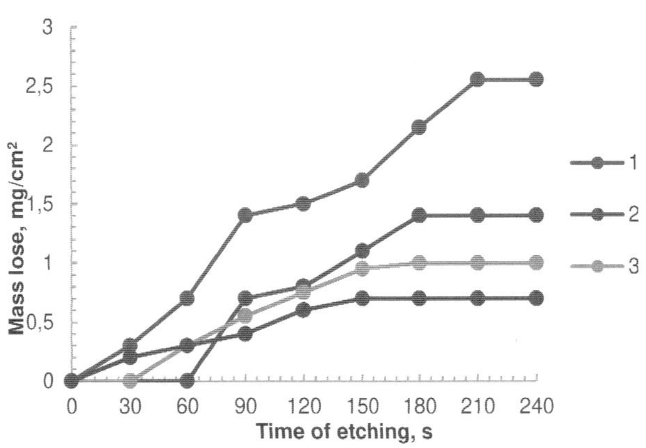

Figure 1. The effect of mass losing during chemical etching of nanoporous alumina (thinness of 1000 nm) at 60°C in mixture 0.6 M H_3PO_4 and 0.2 M H_2CrO_4 (1), 0.6 M H_3PO_4 and 0.4 M H_2CrO_4 (2), 0.3 M H_3PO_4 and 0.2 M H_2CrO_4 (3), 0.3 M H_3PO_4 and 0.4 M H_2CrO_4 (4).

It can be seen (Figure 1), that the raise of concentration of phosphoric acid from 0.3 to 0.6 M increased the value of mass losing of the nanoporous alumina. It may be connected with local dissolution of aluminum and alumina in the bottom of pore in the 0.6 M phosphoric acid. This fact indicated on the good etching ability of phosphoric acid for alumina and remanding aluminum [6]. Analyzing the etching dependences (Figure 1), it should be noted the decrease of mass losing with raising consecration of chromic acid from 0.2 to 0.4 M. Addition of chromic acid goes on to start the partly passivation of aluminum and to lower of etching rate alumina. The end time of etching was indicated by stop of mass changes (Figure 1). The optimal duration of etching of nanoporous alumina (thinness of 1000 nm) was 150 s in mixture of 0.6 M H_3PO_4 and 0.4 M H_2CrO_4. These duration and mixture of etching allowed to safe transparent Al mesh, which had been formed by anodization in 0.3M oxalic acid at voltage of 50 V and temperature of 10°C [6].

The influence of temperature on etching process was exanimated by investigation of transmittance of Al meshes, which were obtained by etching in mixture of 0.6 M H_3PO_4 and 0.4 M H_2CrO_4 during 150 s (Figure 2).

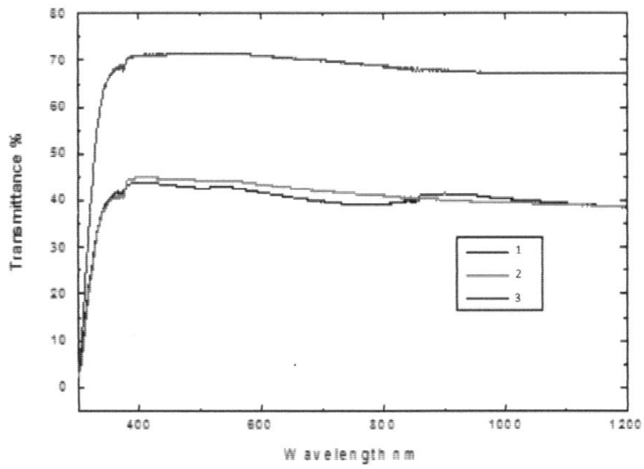

Figure 2. The transmittance spectra of Al nanomeshes obtained by chemical etching in mixture of 0.6 M H_3PO_4 and 0.4 M H_2CrO_4 during 150 s at temperature of 40°C(1), 60°C(2), 80°C(3).

The transmittance value of Al meshes obtained at 40°C and 60°C is lower on average 30-35% then the same value of Al meshes obtained at 80°C. This tendency can be explained by the enhanced dissolution of aluminum oxide and aluminum in 0.6 M H_3PO_4 and 0.4 M H_2CrO_4. This occurs by the increase of the hole diameter of mesh and to remove aluminum from the glass surface (Figure 3).

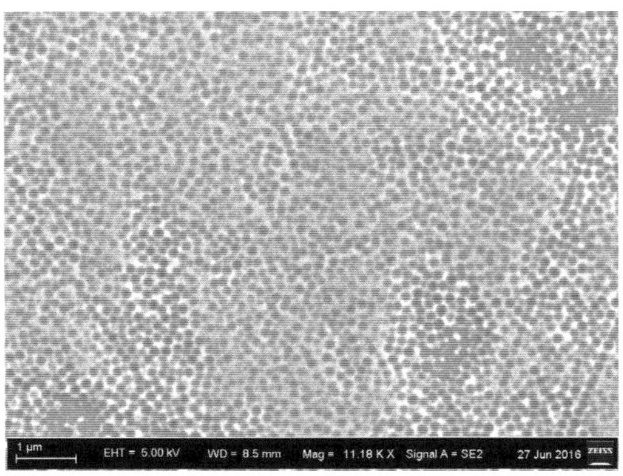

a)

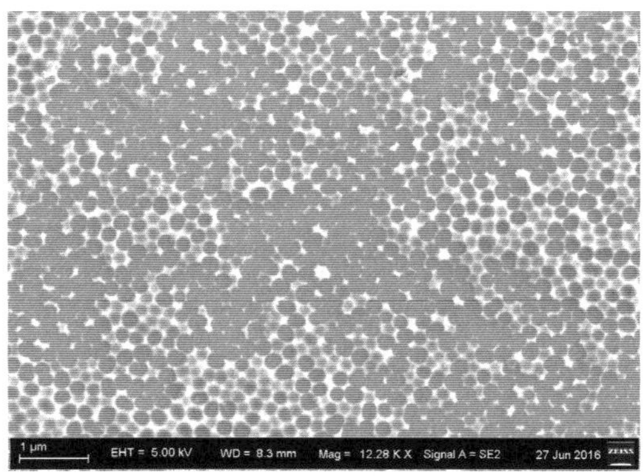

b)

Figure 3. SEM images of transparent Al meshes formed by chemical etching in mixture of 0.6 M H_3PO_4 and 0.4 M H_2CrO_4 during 150 s at temperature of 60°C(a) and 80°C(2).

Defects in hexagonal arrangement of Al mesh were caused by deference in rate of alumina etching in various regions of surface. It should been noted that the little fluctuation of temperature and concentrating of chemical agents in the process of the chemical etching could significantly disorganise hexagonal mesh of aluminium.

The SEM image of transparent Al meshes after anodization by method [6] and etching in mixture of 0.6 M H_3PO_4 and 0.4 M H_2CrO_4 during 150 s at temperature of 60°C are shown in Figure 4.

Figure 4. SEM image of transparent Al meshes after anodization in 0.3 M oxalic acid at 50 V and 10°C and chemical etching in mixture of 0.6 M H_3PO_4 and 0.4 M H_2CrO_4 during 150 s at temperature of 60°C.

The morphology of Al mesh was formed by pyramidal structures (Figure 4), which were organized in hexagonal arrangement onto glass surface (Figure 3 a). The morphology of Al mesh allows developing technique of synthesis of nanostructured functional films with changeable phase composition for chemosensitive layers of new gas sensors. This morphology of Al mesh can provide good alignment properties for LC molecules on the substrate surface and can be one of the promising solutions for the design and production of next generations of displays without indium-tin oxide films [8].

4. Conclusions

Thus, it has been shown that the nanoporous alumina can be used for the production of Al mesh with nanosize pyramidal structures by using chemical etching in the mixture of phosphoric and chromic acid.

The defects in hexagonal arrangement of Al mesh were caused by the little fluctuation of temperature and concentrating of chemical agents in the process of the chemical etching. It should been noted that the increase temperature up to 80°C could to remove aluminum from the glass surface.

These optically transparent Al meshes with good alignment properties for LC molecules can provide creation of new transparent electrodes for display devices without using ITO technology. The morphology of Al mesh allows forming new nanostructured films with changeable phase composition for promising gas sensors and sensory microsystems.

References

[1] Jaguiro P, Stsiapanau A, Hubarevich A, Mukha Y and Smirnov A 2010 *Semicond.Phys. Quant. Electr. & Optoelectr.s* **13** 305
[2] Maeda T and Hiroshima K 2004 *Jpn. J. Appl. Phys.* **43(8A)** 1004
[3] Wang Q, Li Ch, Yuan G and Gu Ch-Zh 2009 *Chinese Phys. Lett.* **26** 86
[4] Sai H, Fujii H, Arafune K, Ohshita Y and Yamaguchi M 2006 *Appl. Phys. Lett.* **88** 201116
[5] Zhylinski V, Bagamazava V, Chernik A, Bezborodov V and Zharski I 2015 *Molec. Crystals and Liquid Crystals* **612** 129
[6] Pismenskaya A S, Chernik A A, Zhylinski V V, Bogomazova N V and Bezborodov V S 2016 *Proceeding of BSTU (in Russian)* **3** 111
[7] Kazarkin B, Stsiapanau A, Zhilinski V et al 2016 *J. of Phys.: Conf. Ser.* **741** 012114
[8] Bezborodov V, Zhylinski V, Chernik A et al 2015 *SID:EuroDisplay* 93

IOP Conf. Series: Materials Science and Engineering **475** (2019) 012031 doi:10.1088/1757-899X/475/1/012031

Polariton-like stimulated excitations in rectifying terahertz GaAs/AlAs double barrier nanostructures driven by nonequilibrity of the resonant-tunneling proces

A A Aleksanyan[1], A L Karuzskii[1], Yu A Mityagin[1], V N Murzin[1] and A V Perestoronin[1]

[1]Lebedev Physical Institute of RAS, Leninsky prospekt 53, Moscow 119991, Russia

Corresponding author's e-mail address: karuz@sci.lebedev.ru

Abstract. We have studied the periodical-in-voltage oscillations in the current-voltage characteristics of GaAs/AlAs double barrier resonant tunneling diode terahertz detectors. The found oscillations of the negative differential conductance are attributed to LO-phonon-branch polariton excitations stimulated by electrons tunneling through the quantum active region in the resonance nanostructure. It is argued that, due to the conduction-band profiles disposition in a manner of the inversion-like nonequilibrium cascade, the tunneling current supplies the inevitable excessive energy. The RTD serves as an internal amplifier of weak electromagnetic signals, provided by the region of negative differential conductance.

1. Introduction

The resonant tunneling diode (RTD) is a quantum-effect semiconductor device using quantum mechanical tunneling which operating principle is based on the resonant tunneling effect. Quantum effects, harmonics generation and high-frequency rectification mechanisms in high-speed RTDs are considered traditionally as rather promising to increase the operation frequency of semiconductor nanoelectronic devices toward the terahertz and subterahertz range. Novel application possibilities develop in resonant tunneling devices exploiting the RTD property attributed to the internal amplification of weak electric signals. RTDs attract the growing interest in the development of sensors for various physical quantities, such as temperature, pressure, light, or microwaves, in the investigation of RTDs with embedded quantum dots operated as single photon detectors and photon counters, or high-gain RTD photodetectors for room temperature telecommunication wavelength light sensing (see [1] and references therein). The advances of RTD devices are resultant of the high internal amplification, low-voltage operation, and supplementary functionality, inherently provided by the region of negative differential conductance (NDC), and as in avalanche diodes, the RTD device performance is tuned by the applied bias voltage. RTD nanostructures are also rather encouraging for possible implementations of the nanoscale thermodynamic cycles [2], similar to the laser cooling [3], for example.

2. Experimental

The computed by a rigorous tunneling theory current-voltage characteristics are less or more idealized functions and comparisons are required with the measured I- V curves, as usual affected by external or internal distortions not considered in the theory. The most basic reasons of the distortions, causing the measured current-voltage characteristic to have the hysteresis, bistability, discontinuities, humps, plane and stepped sections in the NDC, are the excessive series resistance (Rs), parasitic oscillation, charge build-up and longitudinal-optical-phonon (LO) replicas of a tunneling peak [4-7]. There have been several means to extract the parameters of a test resonant tunneling diode accurately: to monitor the oscillation status by measuring the second derivative of I-V curve, to place a resistor in parallel with the RTD, which could stabilize most of test circuits, etc. [6, 7].

These means are used here to extract the reasons of experimentally found periodical-in-voltage features [7, 8] of the NDC. It is argued that, due to the conduction-band profiles disposition in a manner of the inversion-like nonequilibrium cascade (cf. Figure 5 in [7]), the tunneling current supplies the inevitable excessive energy for stimulated excitations in a resonant tunneling process [7]. The found oscillations of the NDC region are considered taking account of the LO-phonon- branch polariton excitations stimulated by tunneling. We found that an RTD serves as a detector of weak subterahertz electromagnetic signals with the internal amplification, inherently provided by the region of negative differential conductance.

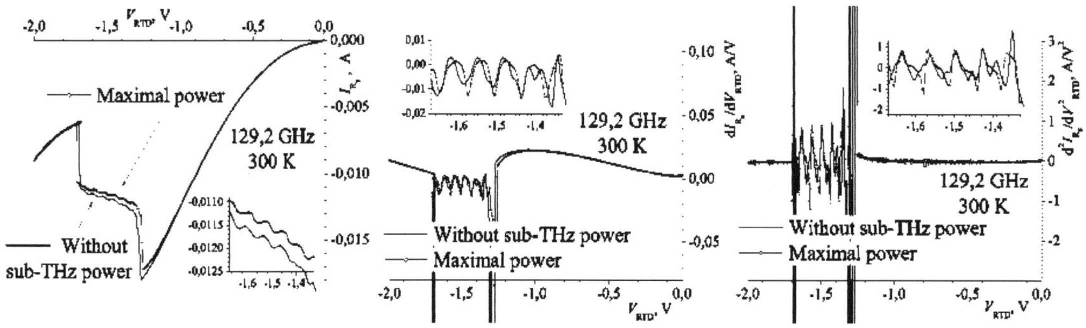

Figure 1. I-V curves (left), first (center) and second (right) derivatives of I-V curves for a double-barrier RTD with and without incident sub-THz power. Maximal incident power is order of 10 /iW [8]. Inserts show more details. T=300 K, active layers of AlAs/GaAs barriers/well 23/45/20 A.

The regular oscillations of the NDC region in the I- V characteristics of an RTD have been detected at room temperature and below [7, 8]. To extract the reasons of these regular features we used here the means discussed in [6,7]: to monitor the oscillation status of the resonant tunneling diode by measuring the second derivative of I- V curve. Information on experimental techniques and RTD working principles, including schematic conduction-band profiles under an applied bias voltage V is shown in Figure 5 in [7] and may be found in [8] and [5, 9], respectively.

The regular oscillations of NDC in the I- V curve are shown in the Figure 1 (left) for a double-barrier RTD with and without an incident sub-THz (0.1292 THz) power at room temperature. First and second derivatives of the I-V curves, obtained from the data points of Figure 1 (left) by machine-routine computations, are presented in the Figure 1 (center, right). Inserts in the figures show more details of the I- V curves and derivatives. From comparisons of the date shown in the figures and those presented in [6, 7] one can conclude the absence of parasitic external oscillations in a circuit with our test RTD. The second derivative curve should show a sharp valley immediately followed by a sharp peak at the lower-voltage point where circuit oscillations would begin. The typical measured derivative curves with diode-curcuit oscillations are shown in Fig. 2(c) in [6]. This feature corresponds to a derivative of the smoothed delta-function in the first derivative of I-V curve, which in turn is a derivative of a step — a negative step of the smoothed theta-function in the I- V curve. This triplet of features may be considered as a classical continuous-media-electrodynamics illustration of a creation-annihilation operator. At the second point corresponding to a higher voltage, as the NDC is smaller negative, oscillations are quenched. The same characteristic of a sharp valley immediately followed by a sharp peak should appear in the second derivative curve at this point also. This pair of the oscillatory features in the second derivative curve can be used to monitor the circuit-oscillation status in the

IOP Conf. Series: Materials Science and Engineering **475** (2019) 012031 doi:10.1088/1757-899X/475/1/012031

measurement circuit and determine the bias voltage range in which circuit oscillations occur. Such pairs of the oscillatory features in the second derivative curve, monitoring the oscillation status in the measurement circuit, can be found, being periodically repeated, in the insert of Figure 1 (right) for a double-barrier RTD with an incident sub-THz power and, contrary, are practically indistinguishable for an RTD without an incident power.

So the observed NDC regular-in-voltage oscillations in the I-V curves are of the intrinsic nature not dependent on the oscillations in an RTD's test circuit. The observed in the Figure 1.

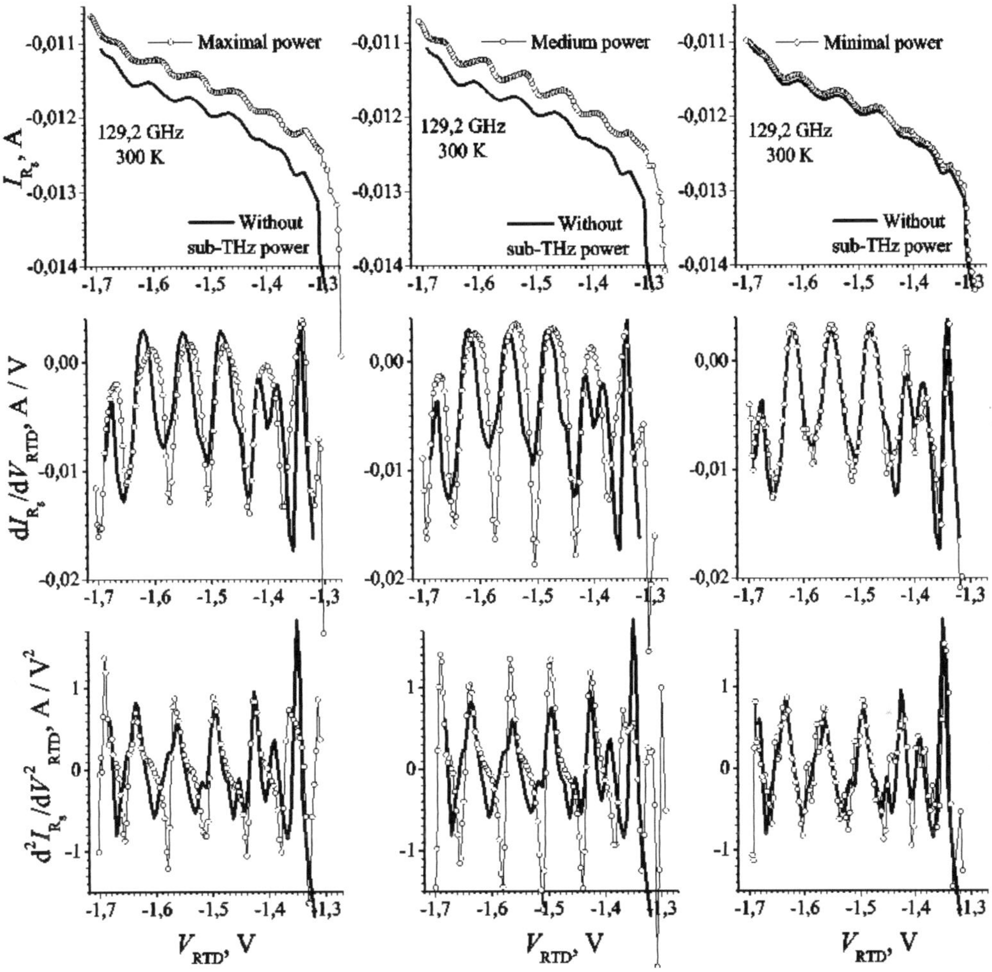

Figure 2. I-V curves (top row), first (middle row) and second (bottom row) derivatives of I-V curves for a double-barrier RTD with and without incident sub-THz power. Maximal incident power is order of 10 /j,W [8]. Data for different levels of incident sub-THz power: maximal (left column), medium (center column) and minimal (right column) are compared with curves in the absence of any incident radiation. T=300 K, active layers of AlAs/GaAs barriers/well 23/45/20 A.

(Left) low averaged value of the oscillatory NDC order of 1/300 fi-1, period of NDC oscillations order of 72 mV corresponding to the energy 72 meV of two LO-phonons under conditions of the slight asymmetry of barriers, number of these periods (7 periods in Figures 1 or 4 periods in [9]) depending on the barrier width (asymmetric of 20 and 23 A in our RTD or symmetric of 12 and 14 A in two RTDs in [9]), and excessive energy of the discussed above cascade inversion [7] provide evidences in support of the resonant tunneling process assisted by a stimulated excitation of LO-phonon-branch polariton replicas. The stimulated LO-phonon-branch polariton excitations in an RTD are provided by the effects of the boundary-dependent local fields in conducting media [7].

3. Experimental results and discussion

These conclusions are supported by comparisons with the demonstrated in Figure 2 data for different values of incident sub-THz power. The triplets of oscillatory features in the I-V curves, first and second derivative curves, monitoring the oscillation status in the measurement circuit, can be found, being periodically repeated for all investigated levels of the weak incident sub-THz power in the Figure 2. These features for the minimal power data are clearly distinct at least for the second derivative curves in the Figure 2 (bottom). Found NDC features cannot be explained by only the ordinary mechanism of the quadrature rectification proportional to a second derivative of the I-V curve d2I/dV2. Contrary, the Ampere/Watt conversion efficiency characteristic of a detector in the form (d2I/dV2)/(dI/dV) [10] allows to conclude that the RTD serves as an internal amplifier of weak electromagnetic signals, provided by the region of negative differential conductance.

4. Conclusion

In conclusions, the found oscillations are considered taking account of the LO-phonon emission stimulated by tunneling of electrons through the quantum active region in the resonance nanostructure. The RTD serves as an internal amplifier of weak electromagnetic signals, provided by the region of negative differential conductance.

Acknowledgments

This research was supported by Grants of RFBR (15-02-09055, 14-02-00658), of Ministry of Education (16.513.11.3079), by Programmes of RAS (24, IV. 12 and III.7)

5. References

[1] Pfenning A, Hartmann F, Langer F et al 2016 *Nanotechnology* **27** 355202
[2] Ziabari A, Zebarjadi M et al 2016 *Reports on Progress in Phys.* **79** 095901
[3] Seletskiy D V, Epstein R and Sheik B M 2016 *Reports on Progress in Phys.* **79** 096401
[4] Goodings C J, Mizuta H and Cleaver J R 1994 *J. Appl. Phys.* **75** 2291
[5] Goldman V J, Tsui D C and Cunningham J E 1987 *Phys. Rev. B* **36** 7635
[6] Bao M and Wang K L 2006 *IEEE Transactions on Electron Devices* **53** 2564
[7] Aleksanyan A A, Karuzskii A L, Kazakov IP et al 2016 *Proc. of SPIE* **10224** 1022400
[8] Aleksanyan A A, Volchkov N A et al 2014 *Vestnik "MIFI" (in Russia)* **3** 671
[9] Suzuki S, Asada M, Teranishi A et al 2010 *Appl. Phys. Lett.* **97** 242102
[10] Dresvyannikov M A, Karuzskii A L, Perestoronin A V et al 2016 *Proc. of SPIE* **10224** 10

Conductivity of superlattices GaAs/AlAs with electrical domains

I V Altukhov[1], S E Dizhur[1,3], M S Kagan[1], N A Khvalkovskiy[1], S K Paprotskiy[1], I S Vasil'evskii[2] and A N Vinichenko[2]

[1]Kotelnikov Institute of Radio Engineering and Electronics of RAS, Mokhovaya 11-7, Moscow 125009, Russia
[2]National Research Nuclear University MEPhI, Kashirskoe shosse, 31, Moscow 115409, Russia
[3]Department of Electrical and Computer Engineering, University of Delaware, 140 Evans Hall, Newark, DE 19716, USA

Corresponding author's e-mail address: paprotskiy@cplire.ru

Abstract. The conductivity of short-period GaAs/AlAs superlattices at the electrical domain formation was studied at room and liquid nitrogen temperatures. The evolution of tunneling current at sweep-up and sweep-down of the bias was investigated. The step-like decrease in current at some threshold voltage was referred to moving domain formation. The current hysteresis and periodic maximums was observed in the electric domain regime. The hysteresis was explained by the changes of electrical domain regimes along with boundary conditions. The origin of current maximum is discussed.

1. Introduction

The study of electron transport in semiconductor superlattices (SLs) has attracted much interest in recent decades [1], mainly in connection with the predicted amplification of Bloch waves, which can made it possible to create tunable sources of THz radiation. Instead of enormous size and power-consuming techniques related to charged beam accelerator for THz generation [2], semiconductor devices are compact and can be rather effective [3]. The main hindrance to the Bloch wave amplification in superlattice structures is the formation of electrical domains owing to the static negative differential conductivity, which eliminates the Bloch gain [4]. However, by analogy with bulk materials such as Gunn diodes, the real part of the impedance of the samples with domains can become negative in a certain frequency range [5-8] and yields high-frequency generation in the corresponding resonator.

To realize THz generation at room temperature, some essential conditions must be filled: a) to obtain short conductivity setting time, which is determined by tunneling in the SL, narrow barriers are required; b) to exclude the thermal excitation of electrons to the higher levels of dimensional quantization and to the conductivity band, narrow quantum wells and a sufficiently large band gap are required; c) fast negative dynamic conductivity in SLs with domains implies a short Maxwell time, therefore high doping level. We have studied previously the effect of a resonant cavity on transport in heavily doped short-period GaAs/AlAs superlattices in the resonant tunneling regime, when the electric domains are formed, and obtained an indication of excitation of THz cavity by the negative resistance of SL with electric domains [9]. In this work, the evolution of current in the presence of electrical domains in GaAs/AlAs SLs was studied.

2. Experimental

The MBE grown GaAs/AlAs SLs consisted of 100 periods of 4 nm GaAs/2 nm AlAs between heavily doped ($\sim 10^{19}$ cm^{-3}) n^+ cap layer and n^+ substrate. GaAs quantum wells (QWs) were doped with $2{\cdot}10^{17}$ cm^{-3} Si. The disk mesa structures were of 7 to 15 mkm diameters. The ring-shaped metal contacts, which simultaneously formed the distributed THz cavity, were deposited near the edge of each disk. A photograph of the structure is given in figure 1. Triangular voltage pulses with a sweep-up times of 0.2 to 30 mks duration applied to the sample allowed us to record current-voltage (I-V) curves for sweep-up and sweep-down of the bias. Measurements were performed at room and liquid nitrogen temperature.

Figure 1. View of the sample with the contact and current lead.

3. Results and discussion

Figure 2 shows the triangular voltage pulse used with respective time dependence of current and reconstructed I-V curve (see figure 3). The step-like decrease (up to 50 %) in current at some threshold voltage is caused by moving domain formation [1]. Several features appeared in I-V curves above the threshold voltages.

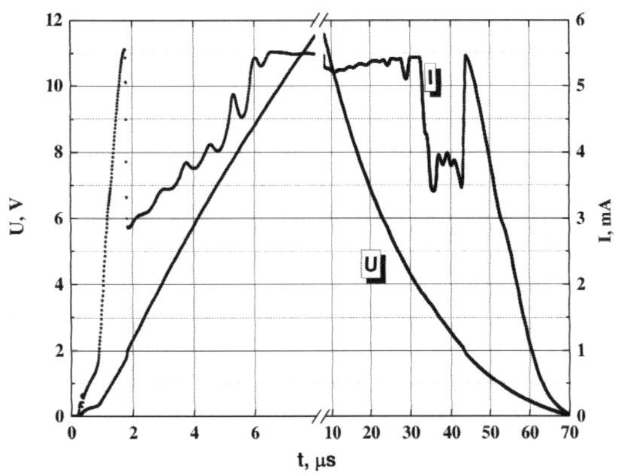

Figure 2. Time dependences of voltage and current.

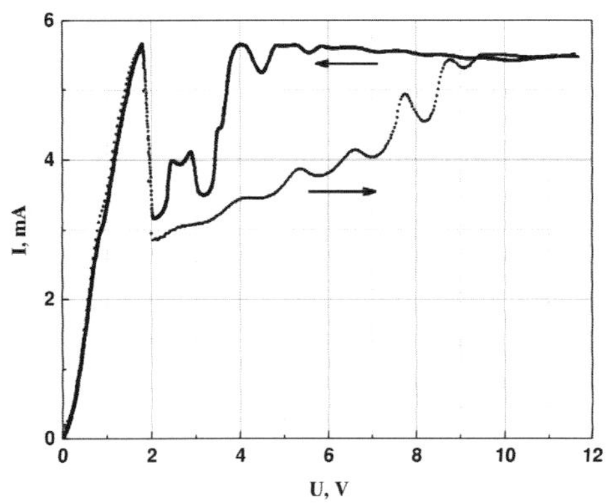

Figure 3. Current-voltage characteristics.

First one is the hysteresis at sweep-up and sweep-down of voltage. The size and shape of hysteresis loop depends on the peak voltage of the pulse. Current-voltage characteristics at different peak voltages are shown in figure 4. The maximal loop observed when the current grew up to its peak value in the I-V curve. The current saturation appears in some voltage range (see curves at 11.7 V in figure 4), being essentially larger at bias sweep-down. The current saturation region on the I-V characteristic at a non-uniform field distribution in the sample, indicates the formation of a static domain and it is swelling up with applied voltage increasing [1]. Thereby we associate this behavior with a transition from moving to static domain at sweeping the bias up and the return at its sweeping down. A similar hysteresis was observed in bulk materials with the Gunn effect and was explained by following: the voltage of a static domain formation can significantly exceed the voltage for maintain it.

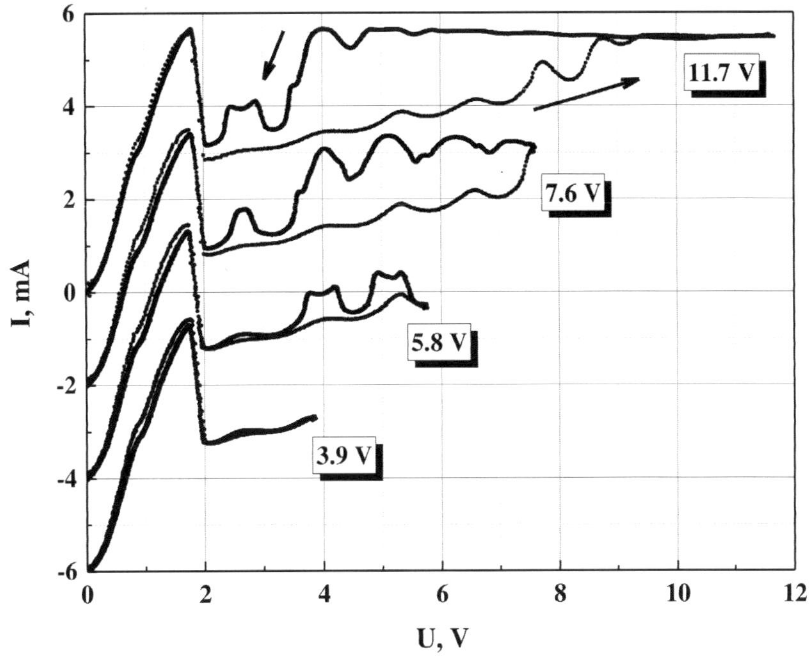

Figure 4. Current-voltage characteristics of GaAs/AlAs SL at forward and backward bias sweep (different peak voltages).

It is possible to estimate the magnitude of the electrical field in and out of the domain from the given I-V characteristics. We have found that outside the domain, the voltage drop on one period of SL is approximately 10 mV, which is noticeably smaller than the width of the miniband. Thus, the overlap of the quantization levels in the neighboring quantum wells has not been removed outside the domain, and the conductivity is determined by miniband tunneling, in contrast to the strong-field region. We can also note the increase of the average current in the travelling domain. This can be associated only with a change of the field in the domain, which can be observed in the case of its triangular shape, similar to domains with the Gann effect in bulk materials of very short-length samples. As the estimations have shown, the charge of domain boundaries is $5*10^{18}$ cm^{-3} to ensure the electric field difference between the regions of high and low fields. While the negatively charged domain wall is formed by free electrons accumulated mostly in one QW, the positive compensating charge is due to ionized donors with the concentration being one order less. Consequently, the field distribution in the domain does not have a flat top, and the domain has a triangular shape.

At the voltage above the threshold (when the travelling domains were formed), the periodic maxima in I-V characteristics were observed. Earlier we discovered the periodic maxima in I-V characteristics of SL structures with cavity at non-resonant tunneling [10] (outside the overlapping of ground confined states in the neighboring QWs). These maxima were referred to the Purcell effect - the enhancement of spontaneous emission rate for optical transitions between confined levels within QWs at resonant frequencies of the cavity. In this case, the positions of the current maxima with respect to the voltage should depend on the main resonant frequency of the cavity. In figure 5 the I-V characteristics for samples with cavity diameters of 7 and 11 mkm are shown. In contrast to the case of non-resonant tunneling, the periodic current maxima at the presence of domains for samples with diameters of 7, 11 and 15 mkm do not shifted and do not correlate with the resonant frequency.

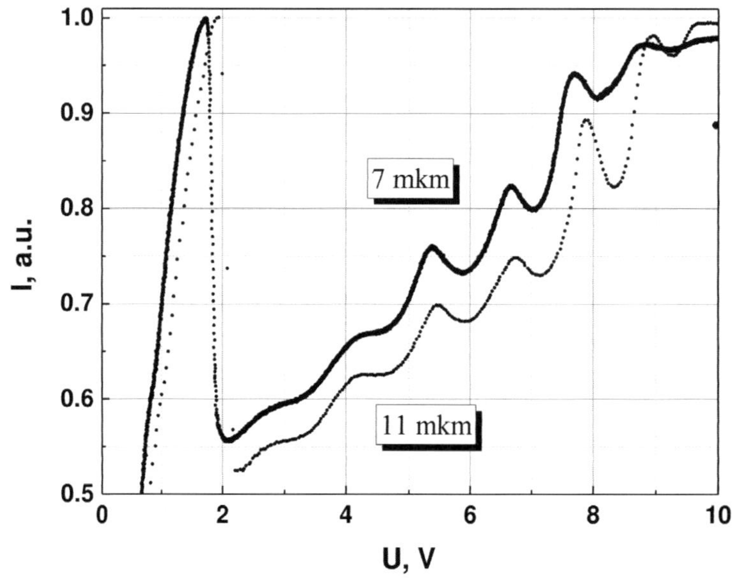

Figure 5. Current-voltage characteristics of GaAs/AlAs SLs with different diameters.

These maxima do not be connected with low-frequency current oscillations too; they are just in I-V curves. To prove this, we changed the duration of bias sweep-up. The time dependences of current for two rising times of the bias pulse are shown in figure 6. The I-V characteristics found from these are identical. However, the periods increased at lowering temperature (see figure 7).

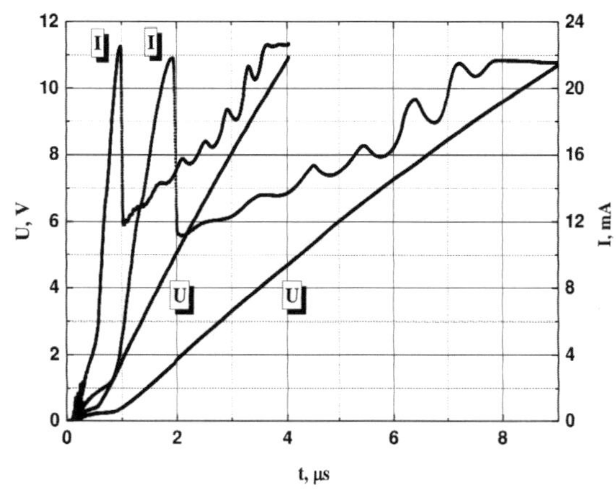

Figure 6. Time dependences of voltage and current at two sweep-up times.

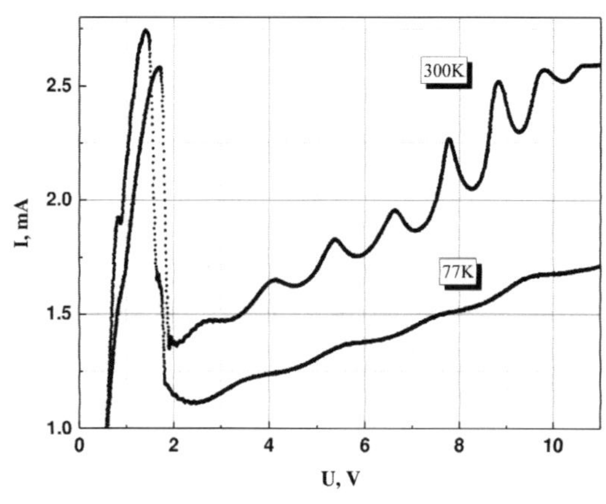

Figure 7. Current-voltage characteristics of GaAs/AlAs SL at different temperatures.

Taking into account the results obtained maxima on I-V are possibly caused by the resonance tunneling of the carriers through several quantum wells in the strong-field region (inside the domain). Note that distinctive features were observed in [11] in I-V characteristics of GaAs/AlGaAs SLs with thin barriers and wide enough QWs (at liquid He temperature), which were attributed to resonant (inter-miniband) tunneling between the QWs located at several SL periods from each other.

4. Conclusion

In the case of short-period superlattices of GaAs/AlAs, the evolution of the current has a rather complicated form, which is associated with the rearrangement of domain regimes and with the resonant tunneling of free carriers in a confined region of the sample (within the domain) through several quantum wells.

The work is supported by RFBR grants 14-02-01062, 16-29-09626, 16-29-03135 and RAS programs "Nanostruc-tures: physics, chemistry, biology, basics of technology" and "New sources of mm and THz radiation".

References

[1] Wacker A 2002 *Phys. Rep.* **1** 357
[2] Ponomarenko A A, Lekomtsev K V et al 2015 *J. of Phys.: Conf. Series* **643(1)** 012119
[3] Chistyakov A A, Kozlovskii K I et al 2016 *J. of Phys.: Conf. Series* **737(1)** 012020
[4] Klappenberger F et al. 2004 *Eur. Phys. J. B* **39** 483
[5] Thim H W 1967 *IEEE Trans. Electron. Dev.* **14** 517
[6] Hakki B W 1967 *J. Appl. Phys.* **38** 808
[7] Zhdanova N G, Kagan M S, Kalashnikov S G 1974 *Sov. Phys. Semicond.* **8(9)** 1731
[8] Altukhov I V et al. 1979 *Sov. Phys. Semicond.* **13** 1148
[9] Altukhov I V, Kagan M S, Dizhur S E et al 2016 *JETP Lett.* **103** 122
[10] Kagan M S, Altukhov I V, Paprotskiy S K et al 2014 *Lithuanian J. Phys.* **54** 50
[11] Andronov A A, Dodin E P, Zinchenko D I, Nozdrin Yu N 2009 *Semiconductors* **43** 228

The deformation-potential scattering and alloy disorder scattering in donor-acceptor pHEMT heterostructures

D Yu Protasov [1, 2], A K Bakarov [1], A I Toropov [1] and K S Zhuravlev [1, 3]

[1]Rzhanov Institute of Semiconductor Physics, Siberian Branch of Russian Academy of Sciences, 13, Lavrentiev avenue, Novosibirsk, Russia, 630090
[2]Novosibirsk State Technical University, 20, K.Marx avenue, Novosibirsk, Russia, 630073
[3]Novosibirsk State University, 2, Pirogov street, Novosibirsk, Russia, 630090

Corresponding author's e-mail address: protasov@isp.nsc.ru

Abstract. The increase in the indium mole fraction in the quantum well of pHEMT heterostructures with donor-acceptor doping causes an increase in the 2DEG concentration from $2.2 \cdot 10^{12}$ cm^{-2} to $3.9 \cdot 10^{12}$ cm^{-2} and a decrease in mobility from 7900 cm^{2}V^{-1}s^{-1} to 7500 cm^{2}V^{-1}s^{-1} at room temperature. The temperature dependencies of 2DEG mobility show that the deformation potential scattering grows at increment of indium mole fraction in quantum well. Absolute values of the deformation potential are approximately two times smaller than the widely used values and increases with increasing of indium mole fraction.

1. Introduction

In pHEMT heterostructures with donor-acceptor doping (DA-pHEMT) the density of two-dimensional electron gas (2DEG) grows sufficiently up to $(4 \div 5) \cdot 10^{12}$ cm^{-2} without appearance of parasitic parallel conductance due to increase of the depth of quantum well (QW) [1]. In the case of a high 2DEG density, two or more subbands in QW are filled. It is known that the interaction between electrons and phonons changes when more than one subband in QW is filled, for example, in $In_{0.53}Ga_{0.47}As/In_{0.52}Al_{0.48}As$ QW the dominant mechanism of electron energy relaxation on the deformation mode of acoustic phonons is replaced by the piezoelectric mode of acoustic phonons; piezoelectric constants differ for each subband [2].

In this paper the scatterings on acoustic phonons and alloy disorder in DA-pHEMT heterostructures with two filled subbands are investigated for the first time. At low temperatures in DA-pHEMT heterostructures the influence of scattering on polar optical phonons can be neglected. In this case the 2DEG mobility is limited by scattering on ionized impurity in δ-layers and acceptor p^+-layers, deformational potential scattering and alloy disorder scattering [3]. Only deformational potential scattering depends on temperature among the mentioned above scattering mechanisms. Therefore, for the investigation of electron-phonon interaction the temperature dependencies of 2DEG mobility were measured in the range $6 \div 77$ K. To study the alloy disorder scattering the mole fraction of indium in QW was varied in the range $0 \div 0.18$.

2. Details of experiments and 2DEG mobility calculations

The DA-pHEMT heterostructures were grown by molecular-beam epitaxy using a Compact 21T machine on (001) GaAs substrate. The buffer of heterostructures comprises 0.4 micrometer-thick GaAs layer and twelve-periods of GaAs/AlAs superlattice. The growth rate was 0.28 nm s^{-1} at the temperature 620 °C and 0.24 nm s^{-1} at the temperature 500÷520 °C for (Al)GaAs layers and InGaAs layers, correspondingly. The growth of δ-layers was carried out at the temperature 530 °C. The layer consequence for DA-pHEMT heterostructures is shown in Figure 1 (without top n+-GaAs contact layer). The nearest to QW δ-layers (filling δ-layers) supplies electrons in QW. The donor concentration in the remote from QW δ-layers was close to the net acceptor concentration for the best compensation. Four samples were grown with the mole fraction of In, y_{In}=0, 0.06, 0.12 and 0.18, respectively. The concentrations of donors (silicon) in δ-layers and acceptors (beryllium) were the same in all samples (see Figure 1).

The density and mobility of 2DEG were determined by Hall measurements at fixed magnetic field of 0.2 T In case of DA-pHEMT heterostructures this method does not give errors [1]. The $\langle 00\bar{1}\rangle$-oriented Hall bars with sizes 50×300 μm^2 were prepared by photolithography method. The Ge/Au/Ni/Au ohmic contacts were formed by electron-beam evaporation and 5-min annealing in hydrogen atmosphere at 420 °C. The samples were mounted at holder with moveable permanent magnet and put down into liquid helium dewar for the measurement of 2DEG mobility at different temperatures.

The 2DEG mobility calculations were performed by using the theoretical model described in our previous paper [3]. The scattering on ionized donors and acceptors, the scattering on deformation potential and alloy-disorder scattering were taken into account. We assumed that the Gaussian distribution with standard deviation σ = 2.5 nm describes the profile of δ-layers. The necessary for 2DEG mobility calculations the heterostructure band diagrams and subband wave functions $\varphi(z)$ were determined using the self-consistent Schrodinger-Poisson solver NextNano.

3. Results and discussion

The obtained values of 2DEG density and mobility at 290 K and 77 K are listed in Table 1.

Table 1. The values of 2DEG density and mobility.

y_{In}	290 K		77 K	
	n, 10^{12} cm^{-2}	μ, $cm^2V^{-1}s^{-1}$	n, 10^{12} cm^{-2}	μ, $cm^2V^{-1}s^{-1}$
0	2.16	7870	2.14	33700
0.06	2.90	7600	2.76	28400
0.12	3.36	7710	3.24	31200
0.18	3.91	7460	3.89	23900

It is seen from the Table 1 that the 2DEG density increases with the increase in the mole fraction of indium in QW. As follow from Table 1 and Figure 1, only for y_{In} =0.18 all donors in the nearest to QW δ-layers are ionized. The 2DEG mobility decreases when y_{In} increments at room and liquid nitrogen temperatures. Unlike to 2DEG density, the mobility changes non-monotonously with y_{In} increase. Such an increase in both the 2DEG density and the degree of ionization is explained by deepening of QW and by the lowering of subband levels relative to Fermi level.

The temperature dependencies of 2DEG mobility are shown in Figure 2. As one can see from this figure, the mobility has linear dependence on temperature at T<68 K and it decreases more rapidly for higher temperatures. Such complicated dependencies can be explained by the influence of scattering on polar optical phonons. The mobility μ_{POP} limited by scattering on polar optical phonons with frequency ω depends on temperature as $\exp(-\hbar\omega/k_BT)$. One can estimate μ_{POP} by extrapolating of linear mobility dependencies to temperature 77 K and using Mattheissen rule. It was obtained that the μ_{POP} values reduce from 1 200 000 cm^{2}V^{-1}s^{-1} for y_{In} =0 to 446 000 cm^{2}V^{-1}s^{-1} for y_{In} =0.18. These values very well compare with the estimation μ_{pop} = 620 000÷540 000 cm^{2}V^{-1}s^{-1} for y_{In} =0.165 from paper [3].

The temperature dependence of 2DEG mobility in DA-pHEMT heterostructures at low temperatures (T<68 K) can be described by the expression:

$$\frac{1}{\mu(T)} = \frac{1}{A \cdot T^{-1}} + \frac{1}{\mu_{const}}, \qquad (1)$$

where $A \cdot T^{1}$ describes the contribution from scattering on deformation mode of acoustic phonons, and μ_{const} – the summary contribution of alloy disorder scattering and ionized impurity scattering. The constants A and μ_{const} were found by the approximation of experimental data by the expression (1) with help of least square method. The approximation results are shown in Figure 2 by lines.

AlGaAs	6 nm
AlGaAs:Be doped	8 nm, 5∗10^{18} cm^{-3}
AlGaAs	7 nm
4 ML GaAs:δ_{Si}	$N_{\delta 1}$, 4∗10^{12} cm^{-2}
SL AlAs/GaAs spacer	2 nm
4 ML GaAs:δ_{Si}	$N_{\delta 2}$, 2∗10^{12} cm^{-2}
SL AlAs/GaAs spacer	3 nm
GaAs smoother	1.5 nm
In$_y$GaAs	14 nm, y=0 - 0.18
GaAs smoother	3 nm
SL AlAs/GaAs spacer	3 nm
4 ML GaAs:δ_{Si}	$N_{\delta 3}$, 2∗10^{12} cm^{-2}
SL AlAs/GaAs spacer	2 nm
4 ML GaAs:δ_{Si}	$N_{\delta 4}$, 3.8∗10^{12} cm^{-2}
AlGaAs	5 nm
AlGaAs:Be	15 nm, 2.5∗10^{18} cm^{-3}
Buffer layers/GaAs substrate	

Figure 1. The layers sequence and doping levels of the DA-pHEMT heterostructure.

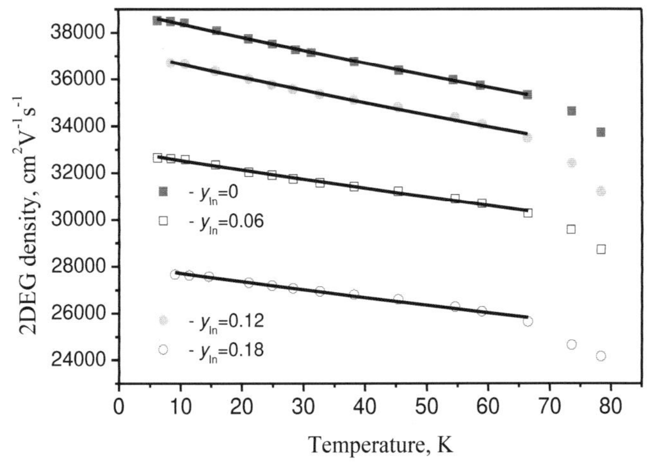

Figure 2. The temperature dependence of 2DEG mobility. Symbols denote the experimental data, lines show the results of approximation.

The obtained values of fitting parameters are listed in Table 2.

Table 2. The values of fitting parameters.

y_{In}	Parameters	
	A, K·cm^{2}V^{-1}s^{-1}	μ_{const}, cm^{2}V^{-1}s^{-1}
0	2.51·10^7	38970
0.06	2.57·10^7	32960
0.12	2.32·10^7	37250
0.18	2.14·10^7	28080

It is seen from the Table 2, the scattering on deformation potential raises at increment of y_{in}. The 2DEG mobility μ_{DP} limited by this scattering mechanisms at 77 K goes down from 326 000 cm^{2}V^{-1}s^{-1} for y_{In} =0 to 278 000 cm^{2}V^{-1}s^{-1} for y_{In} =0.18. The last value in two and half times more than calculated for y_{In} –0.165 in paper [3]. Therefore, the μ_{DP} in paper [3] is underestimated due to wrong choice of deformation potential.

The constant A is inversely proportional to square of deformational potential *D* [3]. Therefore, the dependence *D* on the indium mole fraction in QW was determined by comparing of the experimental dependence $A(y_{in})$ with calculated by using our transport model from [3]. The obtained dependence is shown in Figure 3.

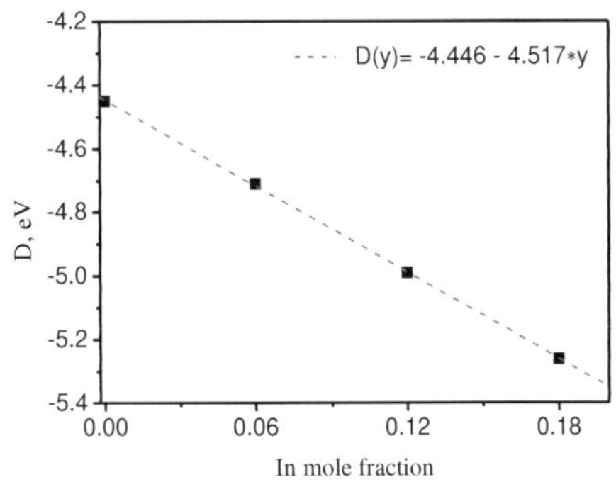

Figure 3. The dependence of deformation potential D on indium mole fraction. Symbols represent the calculated values, dash line shows linear fit.

As one can see, the deformation potential has the linear dependence on y_{In}. But the absolute values of D are less than about twice as widely used values, and grow with increasing y_{In}, whereas they should decrease [4].

Unfortunately, the contribution of the alloy disorder scattering cannot be determined so easy. The mobility limited by alloy disorder scattering μ_{AL} must be reduces with increment of indium mole fraction, because the value of μ_{AL} is inversely proportional to product $y_{In}\times(1- y_{In})$. But a change in the degree of impurity ionization in the δ-layers closest to the QW, with increment y_{In}, also leads to a change in the ionized impurity scattering. This type of scattering dominates in DA-pHEMT at low temperatures, therefore, it is impossible to determine of parameters of weaker scattering mechanism.

4. Conclusion

The 2DEG density in DA-pHEMT heterostructures with same doping level grows proportionally to the indium mole fraction in QW. The 2DEG mobility decreases at increment of y_{In} at room and liquid nitrogen temperatures. The mobility limited by deformation potential scattering μ_{DP} reduces at increment of y_{In} and lies in range 326 000÷278 000 $cm^2V^{-1}s^{-1}$ at 77 K for investigated heterostructures. Such a dependence of the μ_{DP} can be explained if it is assumed that the deformation potential varies according to the following law $D(y)= - 4.446-4.517\times y$ (eV). Due to change of impurity ionization degree and 2DEG density the change in alloy disorder scattering cannot be determined in the samples under study against the background of variability of stronger scattering by ionized donors.

Acknowledgments

This work was supported in part by the Ministry of Education and Science of the Russian Federation in the framework of agreement no. 14.577.21.0250 of 26.09.2017, project identifier code no. RFMEFI577717X0250.

References

[1] Gulyaev D V et al 2016 *J. Phys. D: Appl. Phys.* **49** 095108
[2] Tiras E et al 2001 *Phys. Rev. B* **64** 085301
[3] Protasov D Yu and Zhuravlev K S 2017 *Sol. State Electron.* **129** 66
[4] Vurgaftman I et al 2001 *J. Appl. Phys.* **89** 5815

Influence of silicon donor doping on electron transport in quantum wells AlGaAs/InGaAs/GaAs at different temperatures

D A Safonov[1], A N Vinichenko[1], N I Kargin[1] and I S Vasil'evskii[1]

[1]NRNU MEPhI 31 Kashirskoe shosse, Moscow, Russia

Corresponding author's e-mail address: ivasilevskii@mail.ru

Abstract. Electron transport in single delta-Si doped pseudomorphic quantum wells with increasing donor concentration analyzed in a temperature range 2–300K. Hall effect and Shubnikov-de Haas oscillations studied at low temperatures. Temperature dependences of electron sheet concentration differ: samples with higher doping show significant increase of electron consentration at high temperature, while for the lightly doped it has a minor temperature sensitivity due to the band peculiarities of donor ionization. Electron mobility increases and then decreases at higher electron concentration, although the second quantum well subband remains unpopulated.

1. Introduction

Unlike well-elaborated "classical" high electron mobility heterostructures (HEMT) based on AlGaAs/GaAs quantum well (QW), application-related approach dominates for pseudomorphic structures (PHEMT) with InGaAs QW channel rather than fundamental approach for study of electron processes at high electron density $n_S > 1.5\cdot10^{12}$ cm^{-2}. State-of the art analysis shows an ambiguity of electron mobility dependence on doping concentration in PHEMT quantum wells AlGaAs/InGaAs/GaAs [1-3]. In different papers authors are interpreting either increasing or decreasing behavior of electron mobility with increase of electron concentration in terms of remote ionized impurity scattering mechanism. Thus behavior of μ(n) dependence in PHEMT quantum wells within the wide range of donor concentration haven't a common physical explanation, so the optimization of electrical conductivity and scattering processes in such QW heterostructures is still of great importance.

In PHEMT structures electron concentration can be increased by applying higher donor doping concentration and by double-side doping strategy. Usually in typical pseudomorphic quantum wells second QW subband becomes populated at electron concentration values about $n_s > 2\cdot10^{12}$ cm^{-2} [4], and Coulomb interaction effects lead to significant disturbance of the conduction band profile in the structure. This, in turn, changes electron states' spatial structure and electron scattering conditions. Spacer arrangement also strongly effects the mobility both in HEMT quantum wells and as well in the resonant tunneling structures with electron transport transversal to the structure layers [5].

2. Samples growth and preparation

To study the electron transport peculiarities a series of PHEMT $Al_{0.25}Ga_{0.75}As/In_{0.2}Ga_{0.8}As/GaAs$ samples was grown on semi-insulating GaAs (100) substrates by Riber Compact 21 molecular beam epitaxy. All samples had one-side delta doping with donor concentration N_D values in wide range from

$1.64 \cdot 10^{12}$ cm^{-2} to $15.8 \cdot 10^{12}$ cm^{-2}. For electron transport measurements a Hall-bar mesa structures were made using photolithography and wet etching, with Ni/AuGe/Ni/Au ohmic contacts, annealed at 390 °C. Electron mobility and concentration measurements were carried out at magnetic fields up to 6 T in temperature interval of 2.1 K ÷ 300 K with Cryo-free 404 system.

3. Results and discussion

Increasing silicon doping time in delta layer leads to increase of electron Hall concentration n_H in the structure (fig 1, a). At small N_D values charge carriers concentration in quantum well increases sharply, but at high donor concentration $n_H(N_D)$ dopind efficiency is reduced by ~ 10 times, as it can be seen from the incline change of $n_H(N_D)$ dependence in Fig. 1a. For the further doping increase n_H is weakly increases, despite of increasing N_D by more than 3 times. This might be caused by either worsening of silicon donors ionization efficiency with incomplete incorporating into metallic sites in the lattice or with band-related effects of interplay of donor energy and Fermi level position.

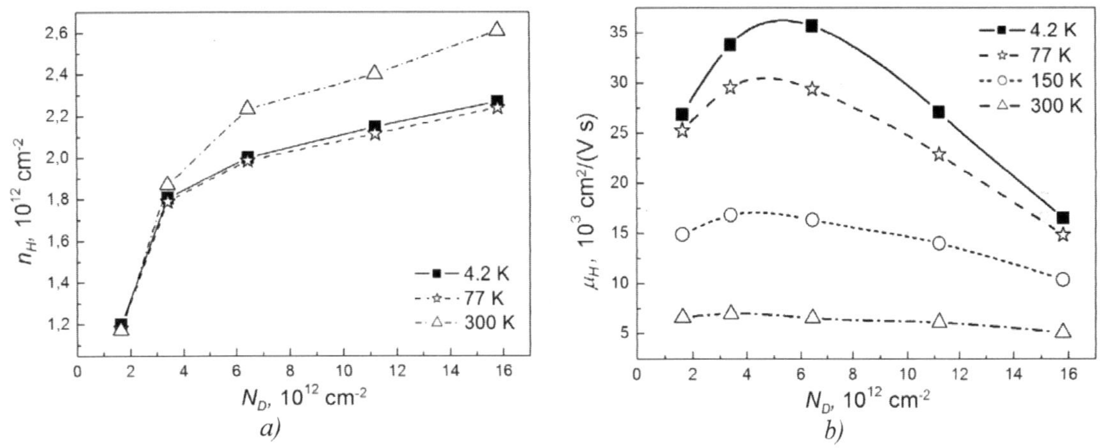

Figure 1. Hall electron concentration (a) and mobility (b) dependences on silicon concentration in delta layer in PHEMT QW samples.

Band profile modelling have shown that increasing of doping impurity concentration in one-side doped PHEMT quantum well of 10.5 nm width formed a quasi-triangular profile of quantum well bottom as a result of appearing strong electric field. Unlike the double side doped PHEMT structures, any considerable population of a second electron subband in QW for the single-doped QWs under study expected only at electron concentration $n_s > 2.7 \cdot 10^{12}$ cm^{-2}.

All samples have been prepared to Hall bar geometry with low resistance alloying contacts [6]. In all samples at low temperatures Shubnikov-de Haas oscillations are observed. Only one oscillation frequency is present in all the samples matching the only populated lower subband in quantum well. Electron concentration in this subband was determined with fast Fourier transform of $\rho_{xx}(1/B)$ dependence and showed a good compliance with hall electron concentration. Besides, one can detect the multiple subbands population by nonlinearity of Hall resistance $\rho_{xy}(B)$. In studied samples $\rho_{xy}(B)$ remained linear even at high temperatures for all the samples, which also indicated the population of only one subband in QW. Thus the decrease of electron mobility with increasing electron concentration in quantum well (figure 1, b) does not related to population of the upper electronic subbands. At relatively low concentration values ($n_s < 1.5 \cdot 10^{12}$ cm^{-2}) electron mobility increases as a result of Fermi-momentum increasing and screening improvement. Then, with further increase of doping mobility decreases because of switching to large angle electron scattering by ionized impurities. This mechanism is caused by an increase of tunnel transparence of wideband $Al_xGa_{1-x}As$ QW barrier for electrons in the ground QW subband due to lowering of conduction band energy with increased silicon donor concentration [5].

Electron mobility shows monotonic decrease with temperature increasing in all considered temperature range (figure 2, a), the most substantial reduction is observed at T > 60 K due to of optical optical phonon scattering mechanism. Temperature dependence of electron mobility character is the same in all samples.

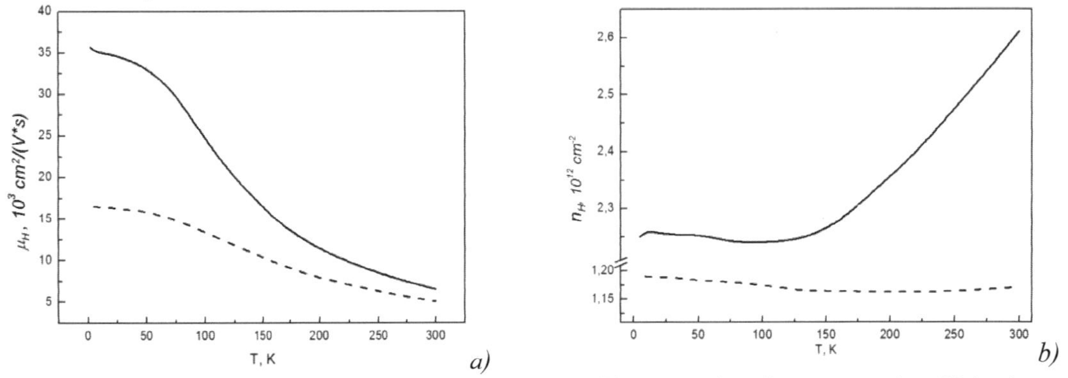

Figure 2. Temperature dependencies of electron mobility (a) and Hall concentration (b) in the PHEMT samples with the lowest and highest Si doping concentration.

Temperature dependencies of Hall electron concentration have a complicated character and differ in various samples (figure 2, b): in samples with moderate doping n_H is weakly changed in a whole temperature range. But in the samples with high $N_D > 3 \cdot 10^{12}$ cm^{-2} a significant n_H increase is observed at higher temperatures. Switching to increasing dependence $n_H(T)$ has different magnitude and takes place at the different temperatures. In sample with lower doping concentration increase of $n_H(T)$ started at T ~ 200 K and has only 2% change of initial value. For the samples with higher doping turn to increasing of $n_H(T)$ dependence occurs at the lower temperatures, and an increase of concentration n_H becomes more pronounced, reaching 15% in the sample with the highest doping.

This distinction is explained by donor ionization temperature dependence with lowering of conduction band bottom potential around delta layer. At lesser doping time ($n_s < 2 \cdot 10^{12}$ cm^{-2}) donor level is located considerably higher than Fermi level and donor activation is high enough and weakly dependent on temperature. With increasing donor concentration conduction band potential at the delta layer begin to lower. In highly doped structures Si: AlGaAs DX-center level is located close to Fermi level and donor ionization is noticeably reduced and became temperature dependent. Temperature of switching to increase of $n_H(T)$ dependence is decreased at higher doping concentration that implies the donor level approaching to Fermi level.

4. Conclusion

PHEMT band profile features define Hall electron concentration temperature dependence: at low doping concentration $n_H(T)$ changes slightly while at high donor concentration $n_H(T)$ increases from the some temperature value, that is decreasing with increasing donor concentration. This phenomena is linked to changes of electrons thermal activation from silicon donor stated with lowering of conduction band bottom potential.

Unlike double side doping, in studied single side delta-doped PHEMT only one subband is populated in the wide range of dopant concentration due to extended asymmetry of quantum well profile driven by high built-in electric field.

Nonmonotonic electron mobility dependence on Hall concentration is observed. Electron mobility increases with increasing donor concentration because of Fermi-momentum increasing. The mobility turns to decrease at the high dopand density because of tunnel penetration enhancement of wideband $Al_xGa_{1-x}As$ barrier for electrons in the quantum well with conduction band energy lowering.

IOP Conf. Series: Materials Science and Engineering **475** (2019) 012034 doi:10.1088/1757-899X/475/1/012034

Acknowledgments
This investigation was supported by the Ministry of Education and Science of the Russian Federation in the framework of the Program of R&D State Orders, project no. 8.3887.2017/PCh.

References

[1] Ohno H et al 1989 *Appl. Phys. Lett.* **54** 36
[2] Greally M G et al 1996 *J. of Appl. Phys.* **79** 8465
[3] Babiński A et al 2000 *Appl. Phys. Lett.* **77** 999
[4] van der Burgt M et al 1995 *Phys. Rev. B* **52** 16
[5] Grishakov K S, Elesin V F 2017 *IEEE Trans. Electr. Dev.* **64(7)** 2963
[5] Vinichenko A N et al 2014 *Semicond.* **48** 12
[6] Vanyukhin K D, Zakharchenko R V et al 2015 *Microelectronics (in Russia)* **44(8)** 564

IOP Conf. Series: Materials Science and Engineering **475** (2019) 012036 doi:10.1088/1757-899X/475/1/012036

Local laser annealing of 3C-SiC film deposited on the silicon substrate by CVD

A V Avramchuk[1], I V Komissarov[1,2], M M Mikhalik[2], V Yu Fominski[1], R I Romanov[1], A O Sultanov[1], N V Siglovaya[1], S M Ryndya, A S Gusev[1], V A Labunov[1,2], N I Kargin[1]

[1]National Research Nuclear University MEPhI (Moscow Engineering Physics Institute), Kashirskoe shosse 31, 115409 Moscow, Russian Federation
[2]Belarusian State University of Informatics and Radioelectronics, Petrusya Brovki, 6, 220013 Minsk, Belarus

Corresponding author's e-mail address: aavramchuck@gmail.com

Abstract. In this work, we try to suite an approach, which concerns of epitaxial graphene growth by laser irradiation of 3C-SiC (111) film deposited on silicon substrate (111) by chemical vapor deposition method. Laser treatment was performed by pulsed 1064 nm laser with 20 Hz repetition rate and 15 ns pulse duration, the fluency was varied 0-1.5 J/cm^2. Raman spectroscopy studies show that for fluence above 0.8 J/cm^2 2D band is noticeable revealing formation of high crystallographic quality graphitic (graphene) film.

1. Introduction

Despite over decade history of graphene intensive study the problem of its integration in electronics remains open. One of the issues which limit the applicability of graphene in modern electronics is the lack of the reliable technology to produce cheap and wafer- scale graphene which electronic properties allow to replace the existing on the market materials. Among the variety of the methods for graphene production CVD and epitaxial growth of graphene on SiC surface are the methods where the most progress has been achieved. Indeed, single-crystals of graphene grown by CVD method possess high charge carrier mobility (up to 16000 $см^2В^{-1}с^{-1}$) [1, 2, 4], compatible with mechanically exfoliated graphene. Moreover, up to cm scale of graphene single-crystal can be grown by CVD [3-4]. However, the need for transfer graphene on target substrate from catalyst surface and problems accompanying this process (wrinkles, charge transfer) [5-6] seriously affect the device performance. On another hand, epitaxial growth of graphene on silicon carbide through thermal sublimation Si from SiC provides direct graphene technology on the isolated substrate. Furthermore, epitaxial graphene grown on different polytypes of SiC substrate is commercially available. It is worth mentioning that except structural quality, the SiC substrates are expensive and size-limited. 3C-SiC film deposited on silicon can act as an alternative to the use of expensive single-crystal substrate of silicon carbide of 4H and 6H polytypes for epitaxial synthesis of graphene. However, because of relatively low melting temperature of silicon (~1400 °C) conventional annealing techniques may not provide optimal conditions for high quality graphene growth [7-8]. The laser annealing could be an alternative technique for ensuring a high local temperature, which could be sufficient to transform silicon carbide into graphene. The possibility of epitaxial growth of graphene by laser annealing on single crystal SiC surface already were reported. S Lee at al grew epitaxial graphene on SiC (0001) using UV laser (KrF

248 nm radiation, λ=248 nm, pulse length 25 ns) in HV conditions ~10^{-6} Torr [9]. M. Hattori et al reported results of patterning of SiC (0001) with 248 nm excimer KrF-laser, pulse duration of 55 ns and repetition rate 100Hz [10]. Recently N K A M Galvão et al grew epitaxial graphene by 10.6μm CO-laser annealing of polycrystalline SiC with 125 W power of laser radiation in atmospheric pressure. They showed that graphene growth as mosaic structure on grains of SiC [11].

In this work we report our structural investigation of 1064 nm pulsed YAG: Nd + laser annealed 3C-SiC film grown by on Si (111) substrate.

2. Sample fabrication

SiC film with ~ 300 nm thickness was grown in cold wall type CVD reactor. Before the growth process, the Si wafer was chemically treated, then annealed at 1100°C in H_2 in CVD chamber. Prior the deposition of silicon carbide, the substrate was pre-carbidized in a 5 cm^3/min C_2H_4 gas for 2 minutes. The silicon carbide film was grown at T=1300 °C using SiH_4 and C_2H_4 precursors for 2 minutes. The flow rate of each gas was 5 cm^3 / min.

The local laser heating of silicon carbide films was carried out in a vacuum chamber at a pressure of 10^{-3} Pa in an Ar gas using 1064 nm YAG: Nd + laser. All experiments were carried out with pulse repetition frequency of 20 Hz, pulse duration of 15 ns. The fluence was varied from 0 to 1.5 J/cm^2. The laser was normally focused on the surface of the film. Finally, 10000 shots were used for each experiment.

The crystalline structure of the silicon carbide film was studied by Ultima IV X-ray diffractometer (CuKα1 λ = 0.154056 nm, tube voltage 40 kV, current 40 mA). A parallel primary beam was applied to reduce the effect of surface roughness on the measurement results. Graphite monochromator was used on diffracted radiation to improve the signal-to-noise ratio. We investigate the structure of the films irradiated with different fluence using Raman spectroscopy (Nanofinder HE (LOTIS TII) spectrometer). The measurements were conducted in the backscattering geometry (spectral resolution of 3 cm^{-1}) with wavelength 473 nm and normal environmental conditions. The surface morphology of the grown films was evaluated by SEM (TESCAN LYRA3 SEM-FIB) and atomic force microscopy (NT-MDT Solver Open).

3. Results and discussion

The diffraction spectra, ω-2θ scan, of 3C-SiC film deposited on Si substrate is shown in figure 1. Three characteristic peaks of 3C-SiC (111), (220) and (222) are visible. The ratio of reflections corresponding to 3C-SiC (111) and 3C-SiC (220) crystals indicates the growth of the (111) orientation of the predominantly single-crystal 3C-SiC film.

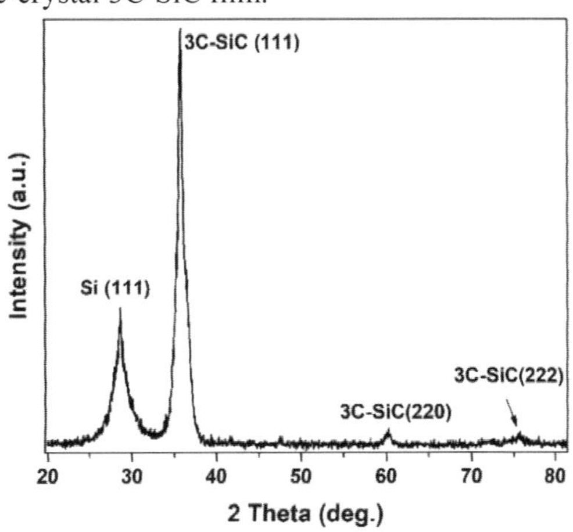

Figure 1. XRD spectra of 3C-SiC film on Si.

In figure 2 we collect the images of as deposited and annealed with 1 J/cm^2 fluence surfaces of 3C-SiC film, a and b respectively.

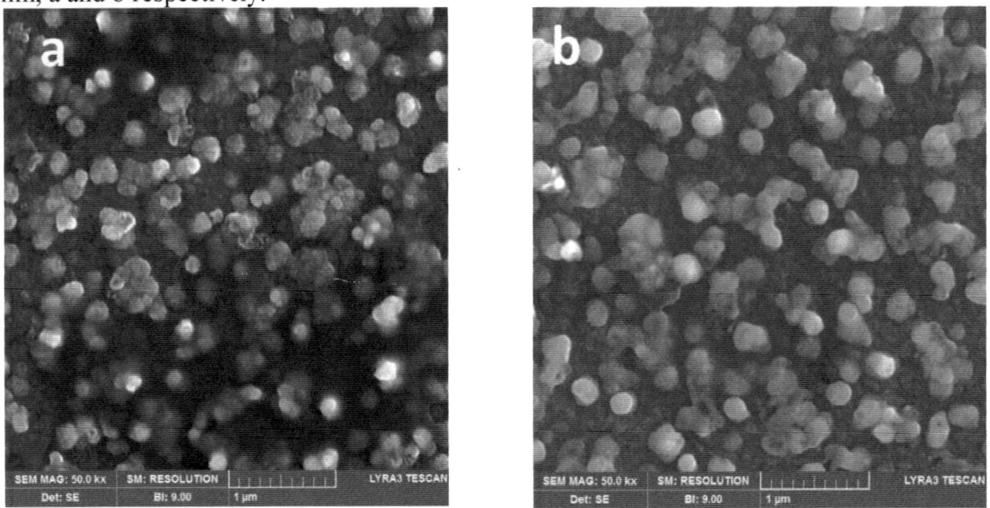

Figure 2. The SEM images of as deposited (a) and annealed with 1 J/cm^2 fluence (b) surfaces of 3C-SiC film.

No significant change in the surface morphology can be noticeable. However, our AFM studies reveal that RMS roughness is decreased from 40 nm for as deposited film to 30 nm for annealed film. This fact already demonstrates significant effect of laser annealing at used conditions on 3C-SiC film.

Our Raman studies (not present here) for the fluence in the 0 - 0.8 J/cm^2 range demonstrate only the changes in 1200 -1700 cm^{-1} domain, which may be associated with formation of disordered graphitic like structure. But for the fluences above 0.8 J/cm^2 an additional peak in 2700-2750 cm^{-1} range associated with 2D band starts to be noticeable. We collect Raman spectra for 3C-SiC surface annealed with laser fluence above 0.85 J/cm2 in figure 3. First, one may notice significant increasing of the intensity of the peak 1450cm^{-1}. Because of the lack of additional structural information, we may only assume the relation if this peak to SiC phase. We believe that intensity increasing of this peak is caused by temperature induced structural changes in SiC. What would be rather in agreement with our assumption for high surface temperature induced by laser annealing. Another significant feature observable in the spectra is the change of the 2D band intensity. Moreover, for the highest fluence we used in our work, FWHM 2D is 31 cm^{-1}. This fact can be explained by formation of much more structurally ordered graphitic like structure on the top of desordered. Unfortunately, strong overlap of G bands (~1603 cm^{-1}) associated with disordered graphite structure and G band from ordered one does not allow to draw any solid conclusion based on G/2D ratio. However, symmetrical Lorentz shape of 2D line together with it position 2731 cm^{-1} support the fact that ordered part of the graphitic layer is fewlayered graphene with turbostratic structure [12]. Nevertheless, the conclusion we made from Raman investigation require crosschecking by other structural techniques. We put our discussion here only as possible scenario.

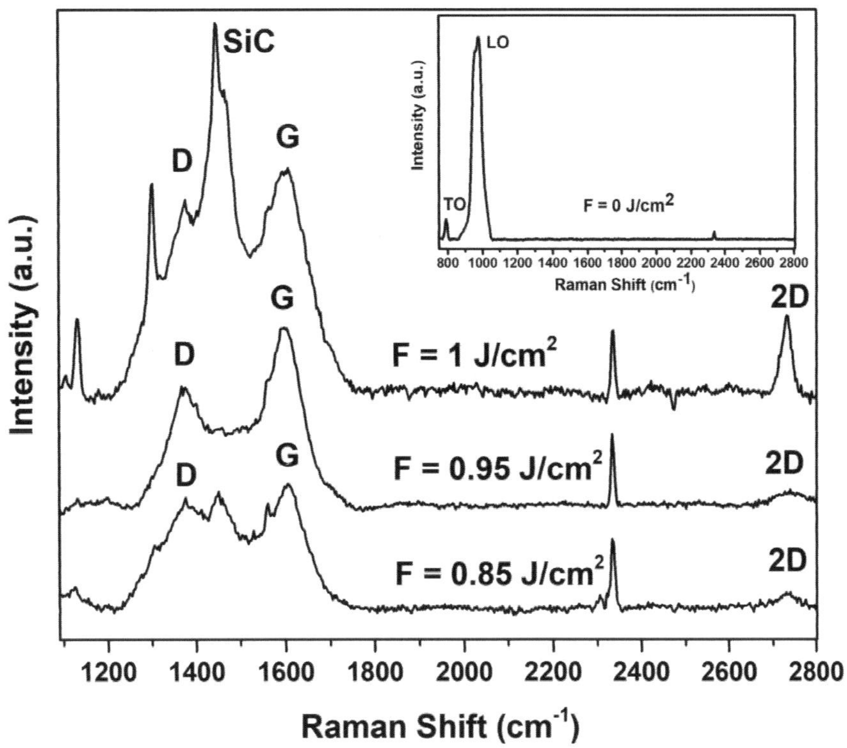

Figure. 3 Raman spectra of irradiated areas of 3C-SiC film with different fluence. The insert shows Raman spectra of as grown 3C-SiC film. LO and TO mark modes for SiC.

4. Conclusion

We demonstrate that local annealing of 3C-SiC film deposited on Si performed by 15 ns pulsed laser with 1064 nm with 20 Hz repetition rate with fluence above 0.8 J/cm^2 leads to formation in Raman spectra of 2D band with FWHM ~31 cm^{-1}. This fact is explained by formation of few layered graphene film with turbostratic structure. The given experimental results show the perspective of the laser annealing for graphene epitaxial growth on surface of 3C-SiC/Si heterostructure.

References

[1] Pirkle A, Chan J, Venugopal A, Hinojos D et al 2011 *Appl. Phys. Lett.* **99** 122108
[2] Li X S, Cai W W, An J H, Kim S, Nah J et al 2009 *Science* **324(5932)** 1312
[3] Lee Y, Bae S, Jang H, Jang S, Zhu S-E, Sim S H et al *Nano Lett.*2010 **10(2)** 490
[4] Zhou H, Yu W J, Liu L, Cheng R, Chen Y et al 2013 *Nature Communications* **4** 2096
[5] J Lu, Y Bao, C L Su, K P Loh 2013 *ACS Nano* **7(10)** 8350
[6] Banhart F, Kotakoski J, Krasheninnikoy A V 2011 *ACS Nano* **5(1)** 26
[7] Ouerghi A, Kahouli A, Lucot D, Portail M, Travers L et al 2010 *Appl. Phys. Lett.* **96** 191910
[8] Gupta B, Notarianni M, Mishra N, Shafiei M, Iacopi F, Motta N 2015 *Carbon* **84** 280
[9] Lee S, Toney M F, Ko W, Randel J C, Jung H J *ACS Nano* 2010 **4(12)** 7524
[10] Hattori M, Ikenoue H, Nakamura D, Furukawa K 2016 *Appl. Phys. Lett.* **108** 093107
[11] Galvão N K A M, Vasconcelos G, Santos M V R et al 2016 *Materials Research* **19(6)** 1329
[12] Faugeras C, Nerrière A, Potemski M 2008 *Appl. Phys. Lett.* **92** 011914